Few people seem to perceive fully as yet that the most far-reaching consequence of the establishment of the common origin of all species is ethical; that it logically involved a re-adjustment of altruistic morals by enlarging as a necessity of rightness the application of what has been called "The Golden Rule" beyond the area of mere mankind to that of the whole animal kingdom.

—Thomas Hardy, The Life and Work of Thomas Hardy

TO MIA, AARON, BEN,

Who prove that the world is enchanted after all

DARWIN LOVES YOU

Natural Selection
And The Re-Enchantment Of The World

George Levine

PRINCETON UNIVERSITY PRESS

PRINCETON AND OXFORD

Copyright © 2006 by Princeton University Press

Published by Princeton University Press, 41 William Street, Princeton, New Jersey 08540

In the United Kingdom: Princeton University Press, 3 Market Place, Woodstock, Oxfordshire OX20 1SY

Requests for permission to reproduce material from this work should be sent to Permissions, Princeton University Press.

All Rights Reserved

Library of Congress Cataloging-in-Publication Data

Levine, George Lewis.

Darwin loves you : natural selection and the re-enchantment of the world / George Levine.

p. cm.

Includes bibliographical references and index.

ISBN-13: 978-0-691-12663-0 (acid-free paper)

ISBN-10: 0-691-12663-1 (acid-free paper)

1. Darwin, Charles, 1809–1882—Influence.   2. Natural selection.   3. Civilization, Secular.   4. Social Darwinism.   I. Title.

QH31.D2L48 2006

576.8'2092—dc22   2006005401

British Library Cataloging-in-Publication Data is available

This book has been composed in Univers and Palatino

Printed on acid-free paper. ∞

pup.princeton.edu

Printed in the United States of America

1 3 5 7 9 10 8 6 4 2

CONTENTS

PREFACE

Some months ago, my son, knowing a great deal about the full range of my beliefs, sharing most of them, and aware that I was still fascinated by Darwin, gave me a bumper sticker that read, "Darwin Loves You." My son is given to an irony and comic cynicism that I have always admired and partly feared, and I was a little uneasy about the obvious aggression that would be entailed in putting the sticker on my car. But there were reasons other than the aggressive and massive public push to religiosity that has so marked the early years of the twenty-first century in America that led me to paste the sticker on after all. I had come to realize that in a perhaps comic, at least ironic way, the bumper sticker was implying something true and important about Darwin that had attracted me to him in the first place and that had continued to attract me after twenty years of study.

It was that realization that led me to shift away from my original intentions in writing this book and to develop them in different directions. I had wanted to consider the strange cultural history of Darwin's scientific theory, the fact that it has been used as support for the most extraordinary variety of cultural, political, and ideological projects. Many who have taken opposed ideological and moral positions have considered themselves true Darwinians. Part of my point was (and remains central to the book as I finally have written it) to defend Darwin from some of the popular conceptions of Darwinism, in particular, from the view that his theory *intrinsically* entails both a radical denial of moral and aesthetic value (because it attempts to explain these phenomena naturalistically) and a simple sanctioning of the worst aspects of dog-eat-dog capitalism.

My overall point was to develop further the argument I have made elsewhere, that scientific and philosophical theories have no *intrinsic* connection with particular political or social positions. Conceding from the start that any philosophical or scientific idea is certain to be marked by the time and place of its conception, the social and political context of its development, I wanted to show how Darwin's ideas were later adapted to many markedly different cultural and political positions. I was convinced by the history of "Darwinism" that when the idea is adopted by other thinkers in other contexts, it is likely to be usable in very different ways, responsive to the newer contexts rather than to Darwin's own. This contingency of ideas means that old ideas in new contexts will take on ideologically different implications. The political and ideological implications of Darwin's ideas (of any philosophical ideas) are not constitutive but contingent. On the one hand, to use my dominant example, a convincing argument has been made for the deep connection between Darwin's thinking and laissez-faire economics, in particular via his reading of Malthus (and this view is thought by many of the best Darwinian scholars to be constitutive rather than contingent),[1] and on the other, there is abundant evidence that his theory has been used to support such incompatible positions as anarchism and socialism.

Yes, for Darwin it may well have been a dog-eat-dog world, but for Kropotkin, to take one example, Darwin's ideas served as a strong theoretical basis for anarchism and mutual aid. Of course, I don't mean to say, then, that Darwin's theory is rightly interpreted as anarchist in orientation, but there can be no doubt that it can be used by smart anarchists, and was so used. The political connections of Darwin's own experience were clearly important to the development of the theory but were not permanently "built in" to the theory itself as it was to float freely through the culture and into future generations. We may find it useful to understand Darwin, as several excellent recent biographies have done, by understanding the particularities of his own social, cultural, and political context, but then, to understand and recognize other versions of Darwin, we would have to understand the particularities of the later theorists' contexts.

The turn my son's bumper sticker provoked in my project follows naturally enough from the consideration of the uses of

Darwin that I had originally intended, and in building the argument of this book toward what now seems to me the bumper sticker's most interesting implications, I have felt obliged to demonstrate (as I try to do, in particular, in the second chapter) how flexible, indeed, Darwin's theory is in cultural interpretation. But the history of interpretations of Darwin is not the history of a series of intellectuals who simply misinterpret him for their own purposes. Rather, virtually all of them legitimately located in his writings arguments that might sanction their own positions (almost, one might add, the way the Bible continues to be mined for ideological possibilities, except that Darwinians seem to be much more careful to think through the whole context of Darwinian thought as they take his theories where they want to go). That is to say, for example, that Kropotkin had strong grounds in finding important evidence that "mutual aid" rather than "nature red in tooth and claw" is the dominant significance of Darwin's theory of the "struggle for existence." Virtually all the interpretations that have been developed are at least partly justified by what Darwin actually wrote.

But working on Darwin in these contentious days, I found it peculiarly difficult to sustain the scholarly detachment that would have been required simply to record with a kind of Weberian disinterest the various more or less legitimate interpretations of Darwin that history has thrown up. Since I do in fact believe that all knowledge is historically contingent but, at the same time, that knowledge is disseminated in ways that will inevitably free it from its initial contingencies (even if only to lock it into later ones), I was not disturbed by the fact that I was finding in Darwin yet another set of cultural implications. My major responsibility, I believed, was to honor what Darwin actually wrote. It is part of the overall argument of this book that any interpretation of Darwin has such a responsibility to Darwin's own words, and to the evidence one can find in his life and work. The fact of contingency does not argue one way or the other for the validity of the theory or the knowledge. There is no way to escape contingency, and while the tendency of criticism has been to assume that once contingency is discovered the validity of the theory is called into question, I want to argue that contingency is the condition for any knowledge and may in fact contribute importantly to the possibility of developing that knowledge at all. Darwin's extraordinarily creative and useful theory developed out of the rigors of his scientific work *and* the pressures of his cultural being.

It is absurd to think that these days one can argue about the implications of Darwin's theory without stumbling into ideological conflict. In America today it is virtually impossible even to use the word "Darwin" without getting a rise out of people. Darwin still makes the front page of the *New York Times* 150 years after the appearance of *On the Origin of Species*. A student in a class I taught that required her to read *On the Origin of Species* told me one day that when she read Darwin on her subway ride she had to cover the book, because not infrequently she was accosted by people hostile to the theory of evolution and to Darwin.

Writing about Darwin and the intellectual history of his ideas, I found myself increasingly disturbed not merely by the usual American resistances to the theory of evolution—it's *only* a theory, it's said, and the school boards run for cover. The blind anti-Darwinism, with its subtler manifestation among the (more or less) scientists of "intelligent design," is now only a symptom of the startling wave of religiosity and, particularly, fundamentalist religiosity, that has entered almost every phase of our public lives. Putting aside the question of widespread ignorance about what Darwin really said, I realized that much of this virtual explosion has come in reaction to what many have argued is a real sense of spiritual vacuity that modern Western society offers its citizens. Reason, science, empirical verification, technological triumphs (and frights) have not been enough to satisfy what seems to be an almost universal longing for "meaning." The world must mean something besides its natural self. Explanation has to satisfy something other than rational curiosity, to point to some significance, some moral ordering, some ultimate justice beyond the disturbing contingencies of the natural world.

I write, however, as someone who found reading Darwin in the first instance a thrilling and enchanting experience. *On the Origin of Species* is one of those books that opened up the world for me, that filled it full of meaning, that inspired and intensified in me a sense of the wonder and enchantment of the natural world. After years of hearing about Darwin, I found the *Origin* to be a book full of personal warmth, of enormous enthusiasm, of wonder and excitement, all these constrained, of course, by a total (and moving) commitment to get the facts right, to build "one long argument" with precision and fairness and openness. The "spiritual vacuity" that, Max Weber had argued, was a consequence of the development of science and scientific explanation, the "disenchantment" of the world that came with a belief that all natural phenomena might someday be explained rationally and naturalistically, hardly seemed Darwinian to me.

As against the life-enhancing energies of Darwinian explanation, I have found the new (largely anti-Darwinian and implicitly antiscientific) religiosity that has followed from the sense of spiritual emptiness dangerous in virtually every way. It has manifested itself clearly enough in the cultism of the late twentieth century and has emerged with political and financial power in the frightening fundamentalisms of the new century, now often cleverly and strategically developed to exploit contemporary means of communication. The fundamentalist disaster of the World Trade Center has evoked a fundamentalist response which, it seems to me at least, is undermining rapidly and violently the structures of democracy and freedom that were built into the American constitution out of Enlightenment aspirations. And flawed as we have discovered those aspirations to be, and inadequate in coming to terms with and understanding non-Western cultures and the spiritual needs of Westerners as well, the fundamentalist alternatives are self-evidently a disaster.

The history of the cultural and ideological uses of Darwinism is partly a history of how, over a century and a half, the culture and a large group of serious intellectuals came to grips with the problem of the potential emptying of the world of meaning that an abstract reading of Darwin's theory might seem to imply. At stake in virtually every cultural exploitation of Darwin's theory was the question of meaning and value: could a naturalistically described world sustain a commitment to moral, aesthetic, and social values? Thus the book I had planned on "the uses of Darwin" morphed quite naturally into a consideration of the fate of "enchantment" in modernity, as science throughout the nineteenth and twentieth centuries expanded the range of naturalistic explanation.

I committed myself then to reconsider, from the perspective of Darwinian naturalism, the narrative of disenchantment, fashioned so powerfully and convincingly by Max Weber, who believed that a rationalized bureaucratic society (like the societies of modern capitalism) given to rational scientific explanation inevitably expels meaning and value from the world. It seems as though just such an expulsion has been at least partially responsible for the explosion of irrationalisms, provincialisms, and violence that is threatening us all. I have not gone mad enough to think that Darwin can save us from all this, but I realized as I continued to think about him and about this new antisecular phenomenon, that the Darwin I first encountered and that I have continued to read now for almost two decades might make an excellent figure around which to build an argument for the possibility of value and meaning in a world gone completely secular. I have committed myself in this book to yet another "use" of Darwin, another interpretation, based on his language, his ideas, and even, to a certain extent, upon his life. Like all other interpretations, all other uses, this one emerges from the contingencies of the historical moment, but, as I have said, I find the fact of contingency not a hindrance but an aid to argument and interpretation.

Surely, Darwin has been a critical figure in the disenchantment narrative, and yet my own experience of reading him has been from the start thoroughly enchanting. In this book I propose to filter through Darwin's thought and work to locate more precisely and more usefully that inspiring and exciting Darwin that I met the first time I read *on the Origin of Species* straight through, and with total attention. His prose, I will be arguing, and his expressed relation to the natural phenomena he describes are—I dare to be sentimental here—manifestations of an intense love of nature and of this world, and it is for that reason that I took my son's somewhat cynical bumper sticker as something far more serious and far more important than, I am sure, its clever inventors, whoever they are, understood it to be.

I am interested, then, in proposing a Darwin, selectively read, as a model of an engaged secularism, as a demonstration that secularism and naturalism need not entail disenchanting aridity. Although it will take some filtering from the historical Darwin, I believe that there is plenty of evidence to justify looking to Darwin as a potential, if not perfect, model for a thoroughly, radically secular but affectively, aesthetically, and morally enchanting vision. The "filtering" I propose is not designed to circumvent his fundamental arguments or to ignore elements in his writing and thought that would seem to be in tension with the model I am trying to represent. I want simply to call attention to certain marked aspects of Darwin's life and writing, to certain personal qualities that flow over into his writing, and particularly into his metaphorical and anthropomorphic strategies.

As opposed to the inadequate alternatives—a religious view of the world that explains it in transcendental terms and makes the experience of this world secondary to that of the transcendent, or a naturalistic view that reduces biology to mere mechanism—I propose here a Darwin who, while absolutely naturalistic in his view of how the world works and of how humans got to be what they now are, finds in nonhuman nature the energy, diversity, beauty, intelligence, and sensibility that might provide a world-friendly alternative to otherworldly values.

A literary attention to his language suggests the possibility of an enchantment that never has to reach beyond nature itself. Of course, in Darwin's world and even in the world of his most extreme, most austerely naturalistic followers, there are "mysteries" enough remaining, although when they are encountered they tend to become "problems." But I am looking here for an attitude, a state of mind and feeling, which, to invert and reexploit Keats's notion of "negative capability," might remain enchanted even in that (ultimately impossible) world in which all mysteries move from problem to resolution. To be enchanted even *without* uncertainties, to be patient in certainty, to find a world potentially explicable in natural terms as thrilling as a world laden with mysteries—that is the naturalistic ideal that I find driving Darwin's life and work. That ideal also drives the effort to write this book.

I am arguing, then, that Darwinian enchantment operates even inside this world in which it is claimed that the spiritual joys of enchantment are accessible only if we can reinstitute a teleology and divine creation. So, beginning with a sense of how diversely Darwin has been used, I have felt myself free to propose yet one more use, no more partial and distorted than most of the others, at least as valid as most that have preceded it, and, if I must say so, much healthier. The argument of this book has thus become an argument for the possibility of an enchantment within a secular world in which science keeps its major explanatory voice, and in which the virtues of rationality are inflected with deep feeling.

But there is sometimes a tension between the affective quality of Darwin's writing and his relation to the natural world and detailed naturalist arguments, and to make an adequate case for my alternative Darwinian narrative, I have needed to address directly just those elements of his thought and life that might seem to sanction the narrative of disenchantment. For one thing, I have had to recognize the strength of the wonderfully historicized Darwin that historians of science like Adrian Desmond, James Moore, Janet Browne, and Robert Young have been giving us over the last two or three decades. But my project of rereading and rewriting the narrative of the relation of Darwin's theory to cultural argument is not, strictly, historical. While I believe that contextualizing scientific thought historically is a critically important activity, that was not the point I was after here. I have sought a way to reinvent Darwin by seeing him inside his historical moment but also by trying to understand how what he was doing transcended the limits of that moment. I have written the book under no illusion that I am representing Darwin in all the complexity of his contingent and historical being, and I am nervous about appearing to have joined the ranks of the hagiographers, so much despised by excellent intellectual historians like Desmond, Moore, and Young. Insofar as this book seems hagiographical it is not doing its work. I try throughout not only to make clear how Darwin's thought and life were deeply embedded in a cultural context whose prejudices and assumptions can often be extremely unattractive, but that there are many unloving and unlovable arguments in his own work. Indeed, part of the point of my argument, and one that I make most directly in the chapter on sexual selection, is that Darwin's ideas were developed out of his thorough absorption in the assumptions and values of his culture. They are unequivocally historical, historical artifacts, and only as a condition of their historicity can I move on to consider how they survive their own moments.

I am not trying to suggest in this book that the Darwin I am focusing on is the complete historical Darwin, or that the virtues I am most enamored of are all there are to Darwin's work. I do want to argue strongly, however, that the Darwin I discuss *is* really there and needs to be attended to. I have tried to filter out of the complex, Victorian-bound, upper-class gentleman's somewhat racist and sexist vision the outlines of a model for what William Connolly calls "nontheistic enchantment." The earth is room enough. Darwin, whatever else, and with all his pains, illnesses, losses, loved the earth and the natural world he gave his life to describing; he found value and meaning in it; he argued that the human sense of value, which he regarded as the world's highest achievement, grew out of the earth earthy, and this genealogy, he believed, did not degrade but ennobled. It is in this sense that I can call myself a true Darwinian and suggest, indeed, that Darwin loves you.

I have written this book and represented the Darwin I most care about because I believe we need a passionate, world-loving secularity, a devotion to the understanding of the workings of nature, and of our own natures. Darwin, with all the flaws and limits historians have been finding, still emerges as one of the great figures in telling us about these things rigorously, honestly, and with feeling. A close reading of Darwin (or at least of parts of Darwin) can put us in touch with the possibility of the blending of reason and feeling, the potential humanity of science, and can put us in touch as well with the wonders of the ordinary movements of nature.

The project, I know, is tinctured with the Victorianism I have studied for most of my adult life. Obviously, I don't think that's a bad thing. Many Victorians, confronted with the shaking news that secularism and science were producing, sought ways to sustain the values that it had been thought only religion could provide. Their project on the whole failed from the (now apparent) naïveté of their faith that they could directly transfer religious values into a secular world—the Positivists, in fact, established a literal church, which George Eliot admired but was too smart to join. In any case, though the project failed, it did so in what I see as a distinguished way. If George Eliot couldn't bring it off, if John Stuart Mill couldn't, then surely I'm not expecting that I'll be able to get very far. But it is a subject too important to let go. Secularity cannot be dead, though it gets strangled daily. Secularity in my own experience has been a condition of a livable nation. I agree entirely with Charles Taylor when he, a convinced Catholic, argues that "secularism is not an optional extra for a modern democracy, we have no choice but to make a go of its only available mode."[2]

But the exigencies of political survival aside, I want to emphasize here another possibility of secularism: the very experience of wonder and joy that Weber suggests is not possible in a "rationalized" world. I have felt, out there in the field watching birds—which I do as often as I can—and there in the study reading Darwin, that the world is saturated with significance and is deeply moving and wonderful (and scary and dangerous). Darwin has taught me a great deal about this and has become for me a model of how to move toward secular enchantment, how to translate the extraordinary joy of a bird sighting, or of dawn against the reddish cliffs of Eastern Arizona, or of the movement of the tides on rocky beaches in Maine, or of the heart-stopping growth of an infant to sight, to recognition, to speech, to balanced movement and personhood, into a realization of what a truly enchanted place this world is, after all.

It matters, I believe, morally and socially—not simply as birdwatchers getting their kicks—that we feel ourselves attached to the world, and in this I have come to value the arguments of Jane Bennett in her fascinating book *The Enchantment of Modern Life*. It matters that we care and that the world in its apparent unintelligent indifference is not alien to us. It matters that we find a way to value the most disparate and improbable and diverse forms of life, and in ways that leave us free to agree that every form of life has a legitimate claim on the goods of this earth. It matters most particularly not because such moments of enchantment liberate us briefly from the pains of modern life but because they allow joy to enter and in so doing help us in our relations to the awfulness of so much that constitutes life around the globe—more than the tedium of rationalized bureaucracies, but the bombs, the slaughters, the rapes, the frighteningly effective lies, the normal difficulties of struggle in the dog-eat-dog world that some people think is Darwin's only gift to us. It matters because it might soften us in our differences, allow us to sacrifice the absoluteness of our perspectives for the compromise entailed in living with difference. We know the whole litany of horrors, and therefore it matters all the more that we find a way to care about the world that seems often so busy destroying itself and threatening us. It matters also, then, that we can recognize in a figure like Darwin someone whose passion for the world drove him to see everything and know as much as he could and tell us what he knew.

"Darwin Loves You" thus becomes quite serious to me, beyond the ironies I'm sure my son was indulging when he gave me, smiling, the bumper sticker. For a book that is otherwise perhaps too academic, it is a risky title, because in its ironies and exaggerations it says something of the greatest importance to citizens of the twenty-first century. I believe it.

In the chapters that follow I move between Darwin's writing and some of the important events of his life. It is important, I believe, in an attempt to understand and recognize the affect built into the language of Darwin's great works, to know something of their historical and biographical context. In talking about Darwin the man, however, I once again am not aiming at some kind of heroizing or hagiographical narrative. As everywhere, I want rather to emphasize the unheroic and culture-bound nature of his life and work. It is a commonplace of virtually all studies of Darwin the man that he was basically a nice guy (although recent biographies have been better at indicating the flaws and limits of his character), and I don't mean to make a big fuss about that relatively unusual fact about a person of world-historical importance. Nevertheless, I have discussed certain aspects of his personal life because my overall argument entails just the possibility that scientific "detachment" and "rationality" can be accompanied by deep feeling and a sense of the value of things. Active scientists are likely to feel insulted by the cultural caricature of the scientist who madly pursues knowledge at any cost, who will risk other lives for the sake of it, and for whom intelligence and moral engagement are utterly separated. Darwin's kind of passion for his subject, his "love" of it, is not very unusual among scientists, who need to have "a feeling for the organism," or a deep personal investment in the natural phenomena they study. But in Darwin's case it is particularly important to emphasize how his relentless pursuit of evidence and arguments for his theory were always a part of a full human and personal investment in it, and always allied to his most profound ethical and cultural commitments.

Although each chapter has a certain independence, I mean each of them to be a part, to cite my favorite author, of "one long argument." In the first chapter, I try to set up the book's overall argument, with particular attention to the idea of "disenchantment." The chapter introduces some of the key social debates about Darwin, attempts to lay out the primary charges against him, and explains the problem of "disenchantment" while proposing a counterargument affirming the possibility of "enchantment" in a scientifically explicable world. In order to build a reasonable case for the Darwin who "loves you," for an enchanted Darwin, I consider in chapter 2 as many of the arguments *against* my reinterpretation of Darwin as I could find, emphasizing just those that find in his work evidence of cultural prejudices or ideas that might lead, for example, to eugenics, to implicit endorsement of the cruelest laissez-faire, or to *Social* Darwinism. I introduce, as well, the perspective of a scientist/critic like Richard Dawkins, who insists, in a way that might provide evidence for Weber's theory, on the full sufficiency of an entirely rationalist, scientific position. Then, in chapter 3, I talk about how Darwin's theories have often been invoked to defend non-laissez-faire positions. This chapter is important, among other things, because it makes clear how sound and interesting, yet ideologically diverse, positions might be sustained by aspects of Darwin's Theory. Chapter 4 continues the discussion with special attention to modern controversies about sociobiology and evolutionary psychology, positions whose ideological implications have frequently been challenged both by scientists and cultural critics. No prior "uses," except those belonging strictly to evolutionary biology, have been so self-consciously based in Darwin's own arguments. These are the boldest attempts thus far to extend Darwin's theory to the explanation of the human condition and human behavior, and thus they are the most obvious and threatening extensions of the movement of "disenchantment." One can't begin to talk reasonably about Darwin's powers of "enchantment" without dealing with these deliberately reductionist interpretations of human mind, behavior, and culture.

Having moved through these chapters that have emphasized the potential disenchanting aspects and uses of Darwin's ideas, I go on in the following chapter, as a kind of preliminary to my case for the Darwin who re-enchants the world, to take a look at a moment in Darwin's biography. In tracing some of the growth of the perspective that would lead to his theory and in emphasizing the perhaps deepest emotional crisis of his life, the death of his ten-year-old daughter Annie, I want to emphasize the connection between his rational, ostensibly unpoetic temperament and the very personal uses of his science. Here I try to intimate the connection between his language, his life, and his theory, with special emphasis on how his scientific instincts do not in the least diminish, or deflect him from, the major crises of his life. Chapter 6 develops this aspect of the argument in relation to one of his major theories. Here I attempt to show that Darwin's almost total participation in his culture's perhaps unreflective conventions and assumptions not only does nothing to undermine the quality and validity of his thought but in effect becomes the condition for the imaginative and intellectual work that produced the theory of sexual selection. It is, then, in the seventh chapter that I can turn my most careful attention to the affective nature of his language, to the ways in which life, feeling, and personal, cultural engagement operate in the work of developing his theory.

And it is in that chapter that I make my fullest case for the enchanting power of Darwin's scientific secularity, finding in Darwin a way of knowing that is deeply human, saturated with value and feeling, and rigorously honest at the same time. I have tried to locate in Darwin's prose something rather more than the literal meanings that continue to be of such importance in the world of science and the worlds of ideology (insofar as these can be separated). Darwin's metaphors, and his prose, his emotional energy, and his pleasures, are as important to my arguments here as his theories. With Gillian Beer, in fact, I believe that the theories, which have been, fairly enough, extracted from the

prose, are deeply informed by those metaphors and by the rhetorical structures in which they are first embodied. The argument
for an enchanted Darwin must build on such things, and I hope
that I have at least made an adequate beginning to that crucial
enterprise, and have managed to turn that facetious and ironic
bumper sticker into something more serious, more credible, and
most of all more useful.

ACKNOWLEDGMENTS

It is hard to think of the world as disenchanted and barren of value when I consider how many remarkable people have mattered to me along the way to this book. I am particularly grateful to Suzy Anger, once more, for advice and support and criticism of my ideas; to Paul Meyer, my dear friend and birding companion, who knows what it is to feel the wonders of nature and birds, who encourages me to think aloud about Darwin, and who cares; to Jim Moore, whose astonishing encyclopedic knowledge of Darwin has frightened me often—if, perhaps, not enough—out of merely sentimental positions and pushed me to more rigorous understanding; to Christopher Herbert, a dogged and brilliant relativist, who knows and loves Victorian literature but, yet better, has come to love the birds, too; to Joe Vining, for whom enchantment, I believe, is still far the other side of Darwin; to the extraordinary and brilliant friends I made during my fellowship at the Bogliasco Foundation where I began putting this book together: Gino Segré, polymath and physicist of distinction; Bettina Hoerlin, who climbs the longest flights of steps unwearyingly and knows the value of irony; Kathryn Davis, who writes like an angel; Eric Zencey, who moves from science to feeling and back with remarkable sensitivity; Anne-Marie Baron, who shops around the world with the skill she applies to Balzac and to the cinema; Martin Bresnick, *landsman* from the Bronx, composer extraordinaire, and *ciambellista* without peer; Lisa Moore, whose genius at the piano leaves me breathless and enchanted; Jenny Jones, who allowed me to share both her glorious discovery of an *upupa* outside her studio window and her imaginative and witty art as well; and Alan Rowlin of the Bogliasco Foundation, whose generosity gives friendship a good name. I want to give my thanks as well to the Bogliasco Foundation itself, for giving me the time, the space, and the Italian air and gardens for work on this book. In addition, other friends have been extremely important to sustaining my passions for Darwin and enchantment: no friend has been more encouraging or helpful than Gerhard Joseph, who listens and speaks sympathetically and wisely and knows as much as anyone about the Victorians and about how to maneuver through the difficulties of life; Carolyn Williams, who urged me to risk the title and my ideas and has always been creatively helpful; Helen Small, who during her stay in America became both friend and intellectual stimulus, and also encouraged me to keep the title; Michael Warner, who first put me on to the work of William Connolly on secularism but, I fear, will find my arguments a little too sentimental; Michael Wolff, a founder of *Victorian Studies*, who has talked fruitfully to me about things Victorian for forty years; Gillian Beer, who first taught me (and a world of readers) to think about how Darwin's rhetoric was intrinsic to his ideas; Jackson Lears, who kept resisting my simplifications in ways that have ultimately enriched my sense of what secularism has been and might be; Cora Kaplan, whose wonderful Darwin conference at Southampton first started me in the directions taken by this book; Jane Bennett, whom I have never met, but who made me believe that my preoccupation with enchantment wasn't entirely mad and that there are after all kindred spirits out there; Rebecca Stott, whose work on Darwin and the barnacles is truly enchanted in just the ways that the Darwin I imagine here should be; Jonathan Smith, a true gentleman and one of the very best literary scholars of Darwin around, who critiqued one of the chapters and has himself written a stunning new book on Darwin; Janet Browne, who read an earlier version of the manuscript with sympathy and a critical eye, raising questions I can't entirely answer, and pointing me toward scholarship that has helped, at least a little, to make my arguments stronger. I owe a particular debt to Vicky Wilson-Schwartz who copy-edited this book in a way that has rescued me from dozens of gaffes; and I have often found her suggestions for rewriting far better than my originals. In quite other ways, I owe much to my son, David, besides the bumper sticker he gave me, not least, of course, his splendid daughter Mia, whose smile is the fullest assurance anyone might need that this world is enchanted after all. My daughter Rachel, and her husband Dale, and their amazing children, Aaron and Ben, compensate for my computer incompetence, constantly puncture my pomposities, and give meaning and value to every aspect of my daily life. And as ever, I owe most to Marge, who has been with me through Bogliasco and years of Darwin and much else, and whose art has grown with our relationship, rich and enchanting.

DARWIN LOVES YOU

CHAPTER 1

Secular Re-enchantment

The gentle gentleman Charles Darwin, who was buried in Westminster Abbey, lives in public consciousness within an adjective describing a brutally competitive and mechanistic world, and as the author of a controversial theory that has made him to many the Antichrist. He has survived not only as the icon of a revolutionary shift in the way we think about origins and humanity but as an unpleasant idea. And for those who think about such things, in extending naturalistic explanation even to human behavior, he is seen as perhaps the most striking embodiment of that scientific rationalism that, in Max Weber's terminology, "disenchanted" the modern world. Evolution by natural selection seems to have removed both meaning and consolation from the world; those who discovered it and who now argue for it often engage in a kind of triumphal rationalism that treads all affective and extramaterial explanation underfoot. It is one thing to believe that science can explain the movement of the stars or even the composition of matter; it is quite another to believe that science can explain human nature itself, and all the disorderly intricacies of human life.

Certainly, Weber's reading of the disenchantment of the world was consistent with the responses of many Victorians to the progress of science. As against the scientific naturalists, T. H. Huxley, John Tyndall, and W. K. Clifford, who exuberantly advertised the power of science to transform the world, W. H. Mallock, among their most brilliant and witty antagonists, noted of the world in a book significantly called *Is Life Worth Living?* that "in a number of ways, whilst we have not been perceiving it, its objective grandeur has been dwindling."[1] Instead of finding that the new knowledge enspirits and enlivens, Mallock claims that "in the last few generations man has been curiously changing."

And the change is the result of too much knowledge, too much reflection. Man "has become a creature looking before and after; and his native hue of resolution has been sicklied over by thought" (19). Mallock's formulation of the Victorian experience can serve as a strong example of Weber's point that the authority of scientific explanation drives meaning and value from the world.

And the Victorian struggle over this problem takes an even starker shape today. One of the more popular scientific books of recent years is called, not immodestly, *How the Mind Works*, and its author, Stephen Pinker, recognizing its immodesty, begins on an uncharacteristic "note of humility" by confessing that "we don't know how the mind works." But, Pinker says, we are on our way, arguing that our understanding of how the mind works has been "upgraded" from a "mystery" to a "problem."[2] And it is precisely the fact that Pinker's project is recognized as a legitimate enterprise of science—the upgrade from mystery to problem anticipates another upgrade to resolution—that, according to Weber, marks modern culture's understanding that science can indeed explain everything. Weber contends that meaning drains out of the world precisely as we come to believe that "if one wished one *could* learn" virtually anything; "there are no mysterious incalculable forces."[3]

There is widespread agreement that this is the case. Pinker's project has deep roots, but in the nineteenth century, particularly in the work of the positivists and scientific naturalists, the enterprise of producing a full scientific description of all phenomena had gained enormous energy. When William James contemplated the project in 1902, he registered a response that confirms Weber's later thesis. "When we read . . . proclamations of the intellect bent on showing the existential conditions of absolutely everything," he asserts with something like contempt, "we feel— quite apart from our legitimate impatience at the somewhat ridiculous swagger of the program . . . menaced and negated in the springs of our innermost life."[4] He talks of "cold-blooded assimilations" that "threaten . . . to undo our soul's vital secrets," and of the "assumption that spiritual value is undone if lowly origin be asserted" (12–13). James's project is to open the way to a recognition of the importance and validity of the religious experience, but to do that he also makes plain the inadequacy for personal and spiritual satisfactions of this scientific "program." He describes, in effect, the condition of disenchantment, about which Weber was to write, and he feels obliged to engage immediately with what is certainly a fundamentally Darwinian project, the explanation of origins in "lowly" terms.

James mocks the pretensions of those who claim to be on their way to describing "the existential conditions of absolutely everything," but that program is not dead. Nor is it self-evident that it's not worth attempting. In effect, it is the program of evolutionary psychology for which Pinker argues, and the outlines of the debate have remained roughly the same over the course of a century, though the technical understanding has changed.

Pinker explains that the mind "is a system of organs of computation designed by natural selection to solve the problems faced by our evolutionary ancestors in their foraging way of life" (x). There is Darwin again, his theory being used here not to explain how species emerge from other species but to explain what is thought to be most distinctive about the human species: mind. It is not a romantic or religious conception that Pinker offers. The mind is a "system," not the seat of the soul; there are problems in its working, but no mysteries. Not only, then, does the use of Darwin imply a disenchanted world, but also (and here is where much modern controversy develops) a world in which morality itself ceases to be a mystery and becomes— again I invoke Pinker—only a problem. As James put it about his contemporary version of evolutionary psychology, the project threatens to undo the soul's vital secrets. The primary problem is that, since Darwin's theory seems to imply that natural selection "acts solely by and for the good of each," that is, it works only on individuals, not on groups or species, it seems impossible to account for "altruism"—the hot issue for sociobiologists and evolutionary psychologists.[5] This is not the place to join the altruism wars of recent theory, but what is clear is that modern uses of Darwin further propagate that sense of him and his work as offering us a world debased because it is explained in terms of lowly origins—as though it were certain that such explanation is somehow degrading. Evolutionary psychology gives us once again a godless nature red in tooth and claw, ruthless competition, survival of the fittest—and now algorithmic social theory and the biologizing of everything human.

When James confronts these arguments, he dismisses the question of "origins" entirely. That is, the quality of an idea, or a work of art, or a person, depends not on its origins but on its effects. It is Jamesian pragmatism carried over into consideration of religion. As for Darwin humankind is not demeaned because we can trace its origins to apelike ancestors, for James religion is not disqualified because we can trace the origins of belief back to some physiological basis. Athough I do not want to dispute that Weber was describing a real phenomenon in registering the dispirited reaction of a culture to the power of rational, naturalistic, scientific explanation, my argument here is Jamesian in that I do not believe that disenchantment follows from naturalistic explanation. James insists on the legitimacy of belief, on willingness to make the bet on the validity of religious experience. My reading of Darwin, in the chapters that follow, points to an entirely secular but similarly satisfactory response. Disenchantment does not follow from a full description of the existential conditions of absolutely everything; and—putting aside the absurdity of the idea that such a description will ever be produced—one does not need to turn to religion to avoid it. Enchantment, of a sort, follows positively, quite naturally, from intense engagement with the entirely secular, and produces—or can produce—a strong equivalent to the condition that James so sensitively describes.[6]

From the outset, Darwin's theories have spurred ideas about the way life is or should be lived. A world of organisms developed from unexplained, apparently random variations, some of which are preserved because of further random alterations in environment—weather, geological transformations, invading species, and so on—seems to yield us a merely chance-driven world, from which the traditional notion of "meaning" has been banished. It was this world against which such clever late Victorians as Samuel Butler and George Bernard Shaw rebelled, and which, I venture to argue, has never been comfortably assimilated by a culture that would yet confess that Darwin was probably right about evolution. The very absence of meaning has seemed to provoke an almost infinite variety of interpretations, and despite Pinker's particular take on natural selection, Darwin has been absorbed into theological as well as atheistical views of nature and life; he has been enlisted for socialism, rampant capitalism, individualism, communal living, natural theology, you name it.

Despite the current upsurge of religious fundamentalisms (itself perhaps a reflex of the "disenchantment" Weber described), continuing and innumerable invocations of Darwin further emphasize the way that "science" has become the most powerfully authoritative language of modernity. Show that an idea is scientific, dress up an actor like a doctor in a television ad, and your claims carry weight. Darwin (along with his popularizers, particularly T. H. Huxley) was a critical figure in the rise of the authority of scientific language. And yet, far from presenting to the culture an unambiguous set of facts about the the origin of species, from the start his arguments provoked alternative interpretations. The problem is not the language's authority but establishing exactly what it is being authoritative about. "Signs," wrote George Eliot in *Middlemarch*, "are small measurable things, but interpretations are illimitable."[7] Darwin's theory is a sign, perhaps not small, but largely measurable. The interpretations to which the theory has been subject are truly illimitable, for it has been invoked for virtually any social or political project.

Scientists have wrangled, and continue to wrangle, over what exactly Darwin meant and what his theory implies, but in the long history of the "development" theory—of descent by modification through natural selection, as Darwin originally termed it—it has been impossible for scholars and social and literary critics to avoid reading his science as ideological.[8] "There is," argues John Durant, "a characteristic tone of moral concern detectable in the writings of almost everyone who is interested in Darwinism at anything beyond the level of the narrowest technicalities."[9] The tendency is uncharacteristic of most other scientific theses, but Darwin and evolution remain hot topics at virtually every level of scientific and cultural discourse, and even at the very technical level they seem to entail that "tone of moral concern."

I

How could they not? As Mary Midgley asserts at the very outset of her essay in Durant's *Darwinism and Divinity*, "Evolution is the creation-myth of our age."[10] It is a myth, not in the sense of being untrue, but in the sense that "it has great symbolic power, independent of its truth" (154). And as such it significantly affects how we think about the world in nonscientific contexts, and how we think about ourselves. The power of Darwin's theory to affect directly all of our lives, manifest in that long series of interpretations and reinterpretations of it and its cultural significance, by scientists, of course, but also by philosophers, by social critics, and by theologians, entails yet further attention. It matters too much to be relegated entirely to science! It is hard to be neutral about Darwinism and hard not to regard every interpretation as rife with ideological and moral significance. A lot is at stake.

Darwin knew it and ducked it as long as he could, but from the very start he understood that he would have to find ways to ease the pain his arguments might inflict on his audience; it is clear that he felt the pain himself and with his various apologetic gestures was not merely defending himself. It is notorious that when he was driven at last to publish his theory, he tried in the first instances to avoid talking about cultural and social im-

plications. And when he began talking about them overtly, in
The Descent of Man, he was not entirely consistent about their human application and usually sought to soften the most disenchanting implications of his ideas. Even in the *Origin* there is a
bathetic and feeble anticipation of the spiritual pain he assumed
most people would feel in confronting a world operating in such
ruthless and mindless ways. The conclusion to the chapter entitled "Struggle for Existence" confronts the horror, recognizing
the need for consolation: "we may console ourselves with the
full belief, that the war of nature is not incessant, that no fear is
felt, that death is generally prompt, and that the vigorous, the
healthy and the happy survive and multiply."[11]

That this won't do is obvious. Others who engaged with his
theory had either to find alternative consolations or reread it in
such a way as to reinsert value and meaning, after all, and because of Darwin's ambiguity, particularly on the question of cultural implications, the possibility of variations in interpretation
were multiplied. As Diane B. Paul has put it, "Darwin's followers found in his ambiguities legitimation for whatever they
favoured: laissez-faire capitalism, certainly, but also liberal reform, anarchism and socialism; colonial conquest, war and patri-
archy, but also anti-imperialism, peace and feminism."[12] Nevertheless, the dominant reading, the one that seems to be implied
by almost every colloquial or journalistic use of "Darwinian,"
takes Darwin's argument as a justification for an unrestrained
capitalist individualism, a mechanically utilitarian ethics, and a
hierarchical structure of races and classes. En route to dispelling
the notion that "disenchantment" is the only possible consequence of Darwin's thought, this book will attempt to modify this
dominant reading.

Entering into these endless (rarely for me tedious, though often deeply annoying) debates, I have my own distinctly non-
neutral moral agenda. Darwin's thinking about nature and the
world remains important; its misuse and abuse have consequences. I hope I will not be misusing Darwin; and I certainly
will not be abusing him. Appropriation of Darwin is, after all,

part of the great tradition of Darwin studies, in which contending philosophers and theorists claim to tell us exactly what it is that Darwin meant and proceed to use "Darwin" to support their own theories and moral programs. Although I understand that every appropriation has its rationale and that it is dangerous to claim that some are "merely ideological" (though in fact I think many are) while others are scientifically objective and value-free, I do believe, and will occasionally argue, that some interpretations are better than others, that some theorists have simply missed Darwin's point or have focused too exclusively on one aspect of a complex argument. Without aiming at an overall synthetic exegesis of what Darwin said or meant to say, I try in this and the chapters that follow to get close to it by attending carefully to one part of what he meant and means that is little attended to and that runs counter to interpretations of his work that focus on its heartless and mechanistic implications.

I will stray some off the beaten paths of technical and literal explication of his views by looking at some of the things he didn't say, by looking at aspects of his life, by considering what others have claimed that he said, and by filtering out from his writing something that he surely meant but didn't say overtly. I will want to be "reading" Darwin with the eye of a literary critic, attending to rhetorical moves in the midst of technical arguments and to the aspects of language that are not literal. The Darwin I will be describing here will not be all of Darwin, by any means, and I suspect some may contend that it's not Darwin at all. Nor will he be entirely "scientific." Although I bear in mind the danger of turning science and Darwinian theory into a kind of religion in its own right, reproducing in mirror image tendencies he struggled throughout his career to resist, I want primarily to argue for the cultural, spiritual, and ethical value of seeing the world with Darwinian eyes. So I am going to appropriate Darwin, in the end, to a set of positions that, I believe, derive directly from things he said or implied; but I do not want to pretend, as I do so, that these are positions that he himself would consistently have supported. But I believe that the Darwin I am

filtering out from his complex, contingent and very Victorian being is important for us, a model of possibilities, rarely until now
addressed as "Darwinian," for the way we might address the
natural world and our society.

I recognize how tricky this enterprise is, how easy it would be
to "filter out" everything I don't like about Darwin and attribute
to him only the things I do like, thus turning him into something
of an intellectual saint utterly removed from any possible reality.
Many years ago, in an essay that attempted to account for the
paucity of literature about "Social Darwinism," except for elaborate and strong arguments dissociating Darwin from his apparent ideologically ugly inheritor, Steven Shapin and Barry Barnes
complained about the developing critical tradition of "purification."[13] It is crucial to my argument that this book not be seen as
a return to the literature of purification, a nostalgic effort to
whitewash Darwin. My argument throughout is not at all that
Darwin is ideologically innocent. How could one say this about
a young man who, making a list of reasons why (or why not) to
marry, comically, but seriously enough, notes, "better than a dog
anyhow"?[14] Rather, I want to build on the reality of his intricate social and cultural involvement with the prejudices of his
moment. I want also to move beyond the question of his "innocence" here, to respond to the fact that he has been used for
many ideological purposes other than those of "Social Darwinism," and to indicate ways in which his work contains the potentiality also—as great literature usually does—for alternative,
humanly satisfying and heuristically promising meaning. My
emphasis throughout will not be on the ways he "transcends"
his culture but on ways his implication in the very texture of his
culture becomes a helpful and creative condition of his work.
The Darwin toward whom I am aiming may, in the end, seem
rather like the Darwin who Shapin and Barnes claim is the object
of the "literature of purification," "an ideally constituted pro-
ducer of knowledge," and my project like the effort they describe and denigrate as "nothing but a way of making Darwin
out as the ideal-type of a modern scientist" (136). But my point is

different: not that Darwin was this "ideal type," but that studying him attentively might help open out the possibility of recon-
sidering *our own* relation to the natural world, our own sense of
value and personal satisfaction, and our own sense of the possibilities of enchantment that—no doubt about it—Darwin's arguments sometimes make it very difficult to sustain.

No doubt, there *is* a clear connection between Darwin's science and rampant, dog-eat-dog capitalism.[15] But a continuing
part of my position will be that, as I have argued recently in another volume, and as Oscar Kensher has convincingly shown,
the connection is not *intrinsic* but contingent; the social consequences do not inevitably accompany the scientific idea wherever it goes.[16] This is not to suggest that Darwin was innocent of
the ideological predispositions historians have increasingly
found in him. No doubt his way of thinking was driven partly
by ideological imperatives. Certainly, outside of *The Origin of
Species,* certainly in *The Descent of Man*, and in other places rather
erratically, he sometimes sounded exactly like the Social Darwinists who followed him. Scholars have been able to trace in
his own public and private writings evidence of the way his theory was linked to particular sets of social and ideological systems fundamental to his moment and his class. The connection
of Darwin's "scientific" theory with Malthus is well known, and
the problem of Darwin's hesitation in publishing his theory has
attracted much attention. Adrian Desmond and James Moore, as
they describe Darwin's progress toward the theory of natural selection, draw the parallel between the theory and politics: "Darwin's biological initiative matched advanced Whig social thinking. That is what made it compelling. At last he had a mechanism
that was compatible with the competitive, free-trading ideals of
the ultra-Whigs."[17] They handle the subject much more directly
in their impressive introduction to a new edition of the *Descent*,
where they trace the expression of Darwin's racist and sexist attitudes. "Science is a messy, socially embedded business, Darwin's particularly so." The historian's responsibility, as they see
it, is to trace the contingencies that radically affect, perhaps even

largely determine, the way the scientist will develop ideas. "In Darwin's case," they say, "Malthusian insights, and middle class mores were central to his theorizing."[18] Although this constitutes a claim rather than a proof, Desmond and Moore make a powerful case for the view that as Darwin struggled to work out the problem of species, his satisfaction with the theory of natural selection had much to do with its ideological compatibility with Whig social theory and politics, and that by the time of the *Descent*, after holding back for many years on direct discussion of the place of the human in the evolutionary scheme he had projected, he brought his deep interest in race—he was passionately antislavery—and his preconceptions about women to the forefront. He was driven, they say, by "abolitionist fervor" (lvii), but equally by assumptions about the superiority of his own class and the superior powers of men as they have, through sexual selection, battled for dominance and possession of the female.

Much current "use" of Darwin exploits confidently the possible laissez-faire connection, and with a kind of tough-minded indignation insists, as Pinker does, on the dominant significance of natural inheritance. One of the better known popularizers of this way of looking at the world is Matt Ridley. Ridley's kind of biologism makes an excellent example for the argument that biologism radically impedes programs for social reform and improvement, giving the sanction of nature to inequalities and injustices that might well be remedied through social intervention. The battle between "nature" and "nurture" has turned nasty, and on the "nature" side one hears in different voices, with different degrees of intensity, the argument, made with some acerbity by Ridley, that "the reason we must not say that people are nasty is that it is true."[19]

Moving from an analysis of the way human nature is conditioned—as scientific and anthropological studies have, to his satisfaction, demonstrated—by self-interest, Ridley goes on to attack cultural theories and theorists who, with a utopian faith in the goodness of people or their governors, propose government intervention as the way to improve the human condition. "So the first thing we should do to create a good society," he says with irony, "is to conceal the truth about humankind's propensity for self-interest, the better to delude our fellows into thinking that they are noble savages inside" (261). With a proper dose of self-irony, Ridley rushes into his own political arguments in a chapter he subtitles, "In Which the Author Suddenly and Rashly Draws Political Lessons." It is at least salutary to have a writer on the "nature" side of the battle admit to the rashness; my point is simply that the political inferences, too often unself-consciously or disingenuously implied, are not inevitable at all. Ridley's argument is that the less we impose governmental control, the more likely we are to get ourselves out of our current moral, social, and political messes. Despising the utopianism of statists, he utopianly takes some of their findings as evidence that nature can pretty much create the good society on its own.[20]

He takes a recognizably Darwinian position, reimagining the process of natural selection in his description—to take a single example from his richly developed arguments—of how the Balinese, on their own, worked out rice farming. Before the intervention of the "Green Revolution in the form of the International Rice Research Institute," the Balinese did very well with their rice. Afterward there was a disaster, and it took further scientific investigation to figure out that before the intervention of "Leviathan" and planning, the Balinese (almost like Darwin's bees, as I understand it), had naturally worked out a system of alternating use of the fields, of the water, and of fallows that avoided the strains imposed by outside regulations and efforts at improvement. Who, asks Ridley rhetorically, was the ingenious person who worked out the traditional Balinese system?

> He was nobody. Order emerges perfectly from chaos not because of the way people are bossed about, but because of the way individuals react rationally to incentives. There is no omniscient priest in the top temple, just the simplest of conceivable habits. All it requires is that each farmer copies any neighbour who does better than he did. . . . All without the slightest hint of central authority. (238)

My point is not to argue about rice in Bali or even about the validity of this implicitly utopian extrapolation of the Darwinian
processes, but only to indicate how easy it is to draw direct political conclusions from "scientific" understanding of human
behavior, and how critical, in any move from Darwinian theory
to cultural theory, is the "nature"/"nurture" distinction (and of
course, implicitly, what constitutes nature and nurture in any
particular situation). The move from is to ought, or from ought
to is, remains problematic. Watching Ridley make that move
provokes, in me at least, a shudder, since he extrapolates his
conclusions about Bali to suggest that state intervention anywhere is destructive. And this even as he has delightedly shown
that people are inherently nasty so that, on his narrative account,
on the way to the utopian solution by nature that his Bali rice
growers had previously achieved, there will have to be a lot of
brutal knocking down and killing, and so on. The less moralized
point is that one could take the same assumptions about the
workings of nature as Ridley does and come up with a radically
different political solution.

The battle over such issues echoes through the history of Darwin's ideas and reputation, and we can think immediately of
T. H. Huxley's insistence, at the end of the "Prolegomena" to
Evolution and Ethics, that however improbable and impossible
the task, humans must resist the "cosmic process" that works
with such brutal amorality through nature. John Stuart Mill's
powerful attack on "nature," in "Nature," as a proper moral
model makes the attempt to identify "virtue" with the "natural"
monstrous and dangerous. But in modern versions of reductionist biologizing, "nature" sneaks back as a moral model, or at
least as a condition that can't be morally attacked or socially addressed. Since there can be no "ought" if nature makes the injunction impossible, what "is" begins to become the moral
norm. To be fair, Ridley is as suspicious as I am of such moves,
but when he argues to politics from nature, lugging in the authority of science to justify his assault on state intervention, he is
making the same sort of move. If nature makes it that way, it's

absurd to try to change it. *The Bell Curve* looms low over the horizon of this sort of thinking, many believe. What, in fact, did Darwin think about the relation of his theory to the work of culture? Does he imply an absolute and permanent connection between human behavior and biological descent or does he allow for the work of culture *upon* the givens of natural selection? Many Darwinians, both left and right, both eugenicists and evolutionists who think eugenics is potentially monstrous, believe that natural selection has been short-circuited by civilization.

Herbert Spencer, in Darwin's own time, took evolutionary theory in the same direction as Ridley does today. Certainly, Darwin's work, as many of the following chapters understand it, was entirely of its moment, a point that the biographical studies of Janet Browne in addition to those of Moore and Desmond have demonstrated.[21] But, I will need to reiterate, it was also brilliantly, doggedly resistant to certain aspects of its moment, original in its capacity to crack old assumptions and take up evolutionary explanation of "species" in ways his predecessors had by and large failed to do. Most originally, of course, Darwin imagined in natural selection the means by which evolution might work. But it is important to remember that "natural selection" did not convince the scientific community until well into the twentieth century. By the second decade of the twentieth century, the idea was almost dead among serious scientists, and only the "new synthesis," a blending, as it were, of Mendel with Darwin, resurrected it. Nevertheless, Desmond argues with impressive historical contextualizing that Darwin's primary fear was that he would be linked through his theory to atheistic and radical materialist revolutionaries who had long since adopted evolution, particularly Lamarckian. Darwin must have "realized," argues Desmond, "how ripe his theory was for exploitation by the extremists."[22] And he feared being connected, as Desmond puts it, with "Dissenting and atheistic lowlife" (413).

Although the Desmond/Moore version of Darwin's fears and class consciousness is persuasive, it certainly is not the whole story; the problem of the history of Darwin's ideas and the uses to which they were put is quite another thing. Despite the ease with which recent cultural criticism has been able to locate acquiescence in the dominant ideology by writers ostensibly dissenting from that ideology, there is no question, first, that they were *in their moments* perceived as dissenting, and second, that even from the perspective of our present, they took positions distinct from those of most of their contemporaries. This applies particularly strongly to Darwin. There is no doubt that Darwin turned evolutionary thought away from what had been its initial direction, among radical antigovernmental thinkers. But the genealogy of evolutionary thought is after all still a genealogy of dissent and resistance to at least some of the varying powers that were, and there is no clearer sign of this than the caution Darwin exercised in publishing his theory and defending it. James Secord has brilliantly and exhaustively demonstrated the way Robert Chambers, in *Vestiges of the Natural History of Creation*, in effect paved the way for Darwin, having changed the conversation about evolution so that it was no longer beyond the pale of serious scientific or polite conversation.[23] Secord's work in developing the evidence for this argument is enormously important, impressive, and convincing, but Darwin's theory, published less than two decades later, remained "revolutionary" in more than trivial ways. Dissenting elements appear as thickly in Darwin's theory, despite his efforts to purge many of them, as do the "ultra-Whig" ones.

Darwin looked for exceptions, for what didn't work, didn't quite make sense—a strategy that ran on the whole against the grain both of the dominant modes of taxonomy and the biblical and natural-theological view of the world as harmoniously designed. Lamarck before him, conceiving of an absolute genealogical continuity in the development of species, had already noted apparent aberrations. But for Darwin, the most interesting aspects of any organism were the "rudiments, echoes of the past, traces of vanished limbs, soldered wing cases, buried teeth—all that conglomeration of useless organs that lie hidden in living bodies like the refuse in a hundred year old attic."[24] Darwin did indeed begin by seeing the world with Paleyan eyes,[25] extremely sensitive to "adaptation," an absolutely key element in his theory and in later uses of it. But in that he allowed himself to focus so centrally on maladaptation, he broke with the tradition of natural theology. As William James was to say, in the light, surely, of Darwin's own writing, "there are in reality infinitely more things 'unadapted' to each other in this world than there are things 'adapted'" (*Varieties*, 478). It was another way of looking at what everyone was seeing, but looking for what didn't work in the dominant explanatory scheme, not for what did. Darwin's ideas gathered their cultural power, not only because they developed out of and reinforced the givens of his moment and the ideological commitments of many who first read him, but because they managed to bring something to the argument that allows them to survive their particular history and feed other, even contradictory, uses. Obviously, the fact of maladaptation was known before Darwin; the drift toward coming to terms with this fact and assimilating it to a coherent story of development and biological life is distinctly post-Darwinian.

"The power of any text," argues Secord, "is not intrinsic, but is always mobilised in particular readings,"[26] and on this view the very idea of "escape" from history is absurd. No idea, I agree, "escapes" history. But it is not wrong to think about the greatness of Darwin's writing or the genius of his conception. It is perhaps true that greatness inheres only in the historical contingency: who is around to read and understand, and under what social circumstances? But it doesn't, then, matter whether greatness is an intrinsic, dare I say, Platonic essence or an achievement limited by the terms of the only history we know. In any practical sense, some writers remain greater than others.

As Derek Attridge has recently argued in a discussion of what it is that constitutes creativity, "the complexity of a cultural field or an idioculture [the sum of cultural forces contained within a single individual] is something we can barely fathom."[27] Originality entails coming to terms with the complexity and dividedness of this fathomless "culture," exerting pressure on its poten-

tial contradictions, recognizing some of its repressions and ex-
clusions (as, for example, the failure of Darwin's contemporaries
to account adequately for aberrations and vestiges). The new
(or the "Other" as Attridge richly analyzes the subject) emerges
through the incoherences and "cracks" in the culture as the artist,
or scientist, more or less consciously recognizes them (25).

It isn't, then, necessary to see Darwin's arguments as in any
way outside of history to recognize their special post-Darwinian
authority. After all, originality can *only* be understood historically and comparatively. On the one hand, we know that most of
Darwin's ideas were already out there for him to assimilate; on
the other, we know that Darwin's thought, as it has "survived"
into the twenty-first century, has been twisted in many ways.
The divisions among Darwinians—most strikingly embodied in
the now famous Dawkins/Gould conflicts[28]—make it absurd to
argue that there is one clear and correct Darwin who has "survived." It is not, as Secord argues, a case of Darwin being "pre-
scient," but it is the case that, however contentious, his ideas
have continued to be useful to scientists and have led to new
ways of thinking about an enormous range of important subjects. Moreover, it is clear that Darwin's intense and persistent
examination of details, of barnacles as well as birds, of caterpil-
lars as well as apes, his work in dissection, his endless question-
ing of colleagues around the world, allowed him to shape the
dominant ideas of his time into new conformations that have
contributed ultimately to a reimagination of the basic myths of
our culture and a rethinking of the relation between biology and
human nature.

Although I will develop this point more fully in chapter 7 as
an aspect of my overall argument, Darwin survives in another
way—as did the great prose writers of the nineteenth century,
like Arnold, Newman, and Pater—because his work is so interesting. Whether we are committed to his idea or not, he represents perhaps the fullest engagement with the natural world
among all Victorian writers and one of the most imaginative conceptions of it that we can find. It is no accident that his ideas and

writings were taken up by so many literary people and transformed into poetry and narrative. To call Darwin's prose "beautiful" may be excessive. He struggled with it always, and there are
signs of that struggle on virtually every page. But it is dazzlingly
imaginative in its metaphorical work; it is rich with "mind experiments" that force readers out of the comfortable niches of their
thought; it implies a vast historical imagination; and perhaps
most important for my arguments in this book, in the precision of
its particular engagements with nature, it implies a passion for it
at least the equivalent of perhaps the most antithetical (to him) of
Victorian writers, the obsessively realistic Ruskin, whose "real-
ism" flowed into the most glorious prose of his time.

Like Ruskin, but perhaps less willingly, Darwin is never un-
controversial. To take a most obvious example, the progressivism that operated in Robert Chambers's *Vestiges of the Natural
History of Creation*, is challenged by, if not entirely absent from,
On the Origin of Species.[29] And yet some of the most interesting
work on Darwin demonstrates that progress lurks in the *Origin*,
and progressivism has, of course, been read back into Darwin
many times. It is central to the meaning of the word "Darwinian" to this day. Darwin was rigorous in insisting that the improvement of species depended on the whims of context, and
yet his arguments are sufficiently ambiguous about the possibilities of constant progress that interpretations can go either way.
Escaping the progressivist tendencies of his own time in constructing his theory, he remained close enough to them to allow
progressivist interpretations to this moment. And when he
came, in *The Descent of Man*, to the development of humans, it is
reasonable to infer that he thought of that development, and
thereafter the development of "civilized" humans, as clearly
progressive.

In any case, the theory has stayed alive because, by and large,
it has worked in almost all of the contexts in which it has been
applied. And whether this is an "intrinsic" quality of the theory
or a condition of the kinds of readings that scientists have been
able to use (barring the period in the early twentieth century be-

fore the "modern synthesis") becomes a kind of metaphysical question. Darwinians now know much that Darwin didn't know, have rejected his notion of blending inheritance and its offshoot, the theory of "pangenesis," have incorporated DNA into the evolutionary scheme to answer (partly) some of the questions Darwin couldn't. Darwinism was, indeed, on the ropes until the Mendelian theory of particulate inheritance replaced Darwin's own mistaken one of blending inheritance.[30] Critics from Fleeming Jenkin in Darwin's own time, to scientists well into the twentieth century asked how individual inherited characteristics could avoid being blended back into the norm of the mass. Darwin had the mechanism wrong; the blending theory, as opposed to the theory of particulate inheritance with the possibility of recessive genes, was vulnerable to Jenkin's finely conceived objections. Darwinism—and belief in natural selection—has thrived since the new synthesis, and most modern Darwinians are far more "Darwinian" than Darwin himself, who always continued to believe that Lamarckian inheritance had at least something to do with speciation. Who is to say, then, that Darwinians now, scientific Darwinians, are truly Darwinian? History is as complicated as Darwin described it, and whatever the ultimate truth of Darwin's arguments, his theories, in some form or other, lie behind the disciplines of evolutionary biology.

The extraordinary multiplicity of interpretations of Darwin's ideas and the abundant and diverse uses of his theory are—at least for the purposes of this book—more interesting as intimations of the theory's power than as evidence for the position that all discourse is endlessly interpretable. The rush, and the persistence, of efforts to make Darwinian theory do ideological work reflect its inescapability and authority. As long as Darwinian theory works in helping us to understand nature, history, ourselves, there will be efforts to assimilate it to strong, ideologically impelled ethical or political programs. Because the theory works, philosophers, scientists, social theorists, politicians all find it necessary to understand it in ways that will support their own particular take on culture, history, and politics.[31] So the multiplicity of interpretations one encounters through history and across cultures is the surest indication of the power of Darwin's theory to survive the limits of his moment and of his first audiences.

Moreover, as I argue throughout this book, at the very moment of their inception, perhaps as a condition of their inception, Darwin's ideas took shape partly in resistance to the conditions that were so important to producing them, partly in exploiting the incoherences in the ultimately "fathomless" variety of his culture. It is a truism of criticism at this moment that every writer, even the greatest, can be understood as in a certain sense an embodiment of cultural forces, an "idioculture," perhaps.[32] But in the end, that truism fails to say anything very specific about a given writer or work unless it is accompanied by meticulously detailed historical research into the peculiar contingencies operating at the moment of the writing: "culture" is simply too big to be reduced to a single set of beliefs and attitudes.

To take only one complex example of the variations within any given "culture" or, for that matter, within any given segment of that culture, when Desmond attempts to analyze the reasons for Darwin's hesitation in publishing his theory, already drafted by 1842, he points out that by then evolutionary and even strictly materialist ideas had achieved in certain contexts and in certain forms thorough respectability; but he also shows that materialism and evolutionary theory were primarily connected with revolutionary thinkers. And yet a tension developed, because some conservative thought was also intricately implicated in evolutionary ideas. So, Desmond points out, "even though the street evolutionists hated the Malthusian weak-to-the-wall thesis, many would still have reveled in the sight of the Anglicans' interfering Deity bound up by law" (412).

Without, then, disputing the connection between "Darwinism" and "ultra-Whig" free-trade liberalism, or suggesting that Darwin did not struggle to make his theory respectable and preserve himself from the dislike of his class and of his fellow scientists, I want to insist that nothing in Darwin's theory *requires* the particular interpretation that leads to Social Darwinism. That is, while Social Darwinism certainly is inferrable from much that Darwin wrote, and I don't mean to "excuse" Darwin from the connection, it has been possible to infer quite different social programs as well. The contingencies of history and the peculiarities of those who interpret ultimately determine the way Darwin's ideas are interpreted.

Robert Young long ago argued that "Darwinism *is* social,"[33] even if he argued the point with enormous impatience that it still *had to be* argued. Young points to many passages, particularly from the *Descent*, that give aid and comfort to future Social Darwinists. And he was unquestionably right. My argument, however, is that Darwin's theory has also given aid and comfort to those, like Kropotkin, for example, who opposed Social Darwinism. Kropotkin or Malthus? Dawkins or Gould? There is evidence for all of them. Whichever social interpretation gets chosen, there is one thing certain about Darwin's theory, and that is the focus of my interest in this book: it unequivocally and unarguably gave support to the idea that the fundamental elements of life, and particularly of human life, are explicable in terms of natural processes. Darwin's theory—though, yes, it has also been put to the service of religion, as I shall be pointing out in later chapters—is a radically secular one. Its primary thrust is that the world can be explained by causes now in operation, that transcendental, supernatural forces do not enter into life. The theory drives toward an explanation of all things, physical and spiritual, by means of natural law.

Spiritual issues are always entangled with biological ones, and that entanglement has continued to the present day, in such enterprises as sociobiology and evolutionary psychology and ecology (of which Darwin is as much the patron saint as of Social Darwinism). But attending to the ways his theory has in fact been interpreted and noticing the ideological swings and possibilities is a first step to the recognition that there is nothing *intrinsic* to the theory that requires the particular political turns that Young rightly emphasized, and there is much in it that suggests alternatives.

This book, then, has two fundamental projects. The first is the point I have been affirming thus far, to discuss some of the many ways Darwin has been invoked and to demonstrate, through discussion of these ways and through interpretation of his texts and aspects of his life, that there is no necessary connection between Darwin's thought and the conventional cultural assumptions about it, and that its cultural and ideological implications and applications are historically contingent.[34] Working within a set of current contingencies, I will, in the following pages, suggest a reading of Darwin that, rather than constraining us to live within dominant ideological systems, can be positively liberating. Stephen Jay Gould, in a discussion of what's wrong with "ultra-Darwinism," as he and Niles Eldridge call a total commitment to adaptationist explanations, argues that "Darwin's system should be viewed as morally liberating, not cosmically depressing."[35]

The second project of this book is, I suppose, both less historical and more personal. I have tried to suggest it already in the preface. It is an attempt, by way of that primary recognition of the contingency of the uses to which Darwin's arguments have been put, to offer yet another way to use him, one designed both to counter the conventional understanding of him as a primary disenchanter of the world and to suggest its reverse, that Darwin's work can be read as contributing to a radical *re*-enchantment of the world. This book, then, is part of a larger project to affirm the possibility and the necessity of that alternative to the sense of the bleak, rationalist world to which I have already alluded, the possibility of what William Connolly calls "nontheistic enchantment."[36]

II

Far, then, from attempting to disentangle Darwin from what Attridge would call his "idioculture," I am eager to see him inside it, partly to make sure that a fair reading of Darwin takes into account the human context of a theory often regarded as ruthlessly inhumane. It is critical to my argument that the very varied "uses of Darwin" be understood as culturally constrained, and since the contexts for other arguments are from different perspectives, different times, different people, the cultural constraints on these arguments will also be different. I intend my contextual reading to serve a second purpose: transformation of the popular word "Darwinian" into an icon of a value-laden secularism.

Weber's narrative of disenchantment is built on his argument that the modern world, beginning with the Protestant reformation but increasingly through Enlightenment secularization, has been bureaucratically "rationalized." One of the most prominent forces in this increasing rationalizing (and routinizing) of life has been and is the work of science. For Weber, the bureaucratizing of the world, the impersonality, routinization, and mechanization that mark the efficient and rationally organized structures upon which modern Western societies depend, leads to the replacement of the "cultivated man" by "the specialist type of man" (Gerth and Mills, 243). Although Weber is rigorously, perhaps excessively, careful to avoid value judgments in his work, there is no doubt that he is registering a loss. His narrative of the rationalization of modern society is precisely the narrative of disenchantment, the narrative of the disappearance of the sacred and mysterious from this world. Disenchantment, Weber insists, consistently affirms that without magic, without God, without teleology, enchantment is purged from the world, and, with it, the world's meaning and the world's value. In the "intellectualization of the world," Weber says, "scientific progress is a fraction, the most important fraction" (Gerth and Mills, 139). Weber's narrative of disenchantment leaves only these options: either a value-laden world infused with transcendental meaning, or an amoral world from which all value is drained as it is subjected to scientific investigation.

Who, more than Darwin, subjected nature to naturalistic and materialistic explanation? Who was more important in making the human the subject of scientific investigation and explanation? Putting Darwin's work in the context of Weber's narrative,

I recognize how neatly, at first reading at least, Darwin fits. As Daniel Dennett puts it (and I will discuss his views of the matter in chapter 7), the theory of natural selection can be recognized as an "algorithm," a kind of recipe that works with mechanical certainty but on the basis of mindless developments in the material world. The resistance to Darwin that is growing so powerful and effective in contemporary American society is clearly a rejection of the mechanistic impersonality such a view entails. A mindless algorithm replaces an intelligent creator, and the world empties out of meaning.

My own argument throughout this book is based on a triple vision of Weber's "disenchantment" narrative. First, that Weber is right to recognize the power of "rationalization" to demoralize. Second, that the demoralization produced by rationalization is far from universal. And third, that some alternative to traditional reliance on the transcendent and the teleological to sustain value and give meaning to life is a genuine human need.

My object in rethinking Darwin here is to attend to this central aspect of human experience and belief, and thus to propose an alternative Darwinian world: a world "bereft" of transcendental spirit that is yet laden with value and entails a deeply emotional, a "visceral," response to the workings of nature. As he tried to wrest the world from theological to scientific explanation, Darwin did not, I want to argue, wrest it away from value or from the kinds of consolations that religion has for the most part been called upon to provide. The very act of trying to understand the world materially and naturalistically entailed right from the outset of his career the attitude of wonder that is so central, on all accounts, to the experience of enchantment.

Natural theology, the explanation of "adaptation" that Darwin was determined to displace, is a kind of theodicy: it justifies the ways of God to man by showing that the world answers, as the Bridgewater Treatises were to formulate it, to "the Power, Wisdom, and Goodness of God." It demonstrates that God must exist and that a careful look at his creation will show that the evil within it is part of a loving plan for mankind. Darwin's theory, on the other hand, is what I'll call a geodicy, a demonstration that the world in all of its wonderful diversity and stark contrasts makes sense entirely on its own terms, although without taking the satisfactions of human desire as its primary goal. It does not prove the world's love of mankind (far from it); it is not built for human benefit but rather includes humans within its family. The world is thick with value because human perception is intrinsically a mode of feeling as well. Connolly, citing the work of Joseph LeDoux, talks of "the several human brains involved in our thought-imbued emotional life" (28); our capacity to think is never separate from the activity of feeling. Darwin's theory does not pretend to avert the evil that even the most enchanted among us must experience and confront, but in the midst of the clear-eyed, often pained perception of natural processes, it sustains the enchantment of the material world.

In such a world, enchantment is not easy or constant. It is never worth having without an awareness, as Jane Bennett puts it, of "the world's often tragic complexity,"[37] which can never be justified. But it allows for the possibility (in fact, I would argue, it is a condition of the possibility) of caring for, or loving the world, even with all its "tragic complexity." The importance of secular enchantment is nicely suggested by Connolly: "attachment to the world," he says, "provides an invaluable source for participation in the politics of social justice" (16). As we follow Darwin's tough geodicy we find ourselves in a world of wonders, a world worth loving; we become participants and observers in a life larger than any of us, and more meaningful.

Without then falling into the camp of hagiographers who find no faults in Darwin's ideas or life, I want to filter out of their complexities and contradictions a kinder, gentler Darwin. Of course, Darwin saw himself as a scientist and aspired to the condition of scientific work that he read about and admired in John Herschel's *Preliminary Discourse on the Study of Natural Philosophy* (1830). But that book itself was thick with a Romantic passion for knowledge, a sense of the divine significance and richness of the natural world. And so Herschel claims that the scientist, "accustomed to trace the operation of general causes, and the exemplification of general laws, in circumstances where the uninformed and unenquiring eye perceives neither novelty nor beauty, . . . walks in the midst of worders."[38] Informed and enquiring scientists, then, in Herschel's view, have a virtually clerical position, just because they are so much more aware of the "wonders" of the natural world. "The uninformed and un-enquiring eye" fails to recognize the beauty that is all about and that a refined and intense attention will reveal. For Herschel as, I claim, for Darwin, the scientific attitude does not merely rationalize the world, explain it away, but opens it up, and makes its wonders available where they have been hidden from less inquiring consciousnesses. For Darwin, the project of establishing the theory of evolution by natural selection was not so much the affirmation of a mindless and godless world, as the revelation that we walk in the midst of wonders; it was an act of loving engagement with the natural world that allows and fosters, even without gods and traditional forms of consolation, enchantment.

I recognize that the project of enlisting Darwin on the side of the angels is a tricky business. There is an enormous danger here of merely sentimentalizing both him and nature, and of diverting attention from the kinds of anomie, mere instrumentalism, and social demoralization and exploitation that have indeed been part of the experience of modernity. The position I am taking up is, perhaps, consistent with the sort of relation to nature that Charles Taylor describes as one aspect in the development of the modern sense of identity; and of course, the position is limited and in some ways incomplete. Taylor contrasts the fundamentally "instrumental" relation to nature that is associated with Enlightenment thought with a Rousseauvian and Romantic one. "Efficacy," he explains, "is valued" in the instrumental view of nature, "as the fruit and sign of rational control," and that control was taken as "a realization of man's spiritual dimension."[39] The Romantic relation to nature constitutes a critique of the instrumental one, of its exploitation of man and nature, its denial of community (276). But, as Taylor points out, the Romantic relation to nature tends to be an aspect of private rather than public life, and the instrumental one, consistent in many respects with Weber's description of the world disenchanted by science, is fundamental to public life. Both, Taylor claims, imply a fundamental valuing of something beyond the merely material. But both, in different ways, move toward the spiritual only by means of nature.

The Romantic relation to nature, as Taylor describes it, approximates the sort of re-enchantment I suggest is facilitated by Darwin, or perhaps by my reading of Darwin:

> discerning the demands of nature involves identifying my true sentiments, setting aside the false (because unnatural, heteronomous) passions. It requires a kind of intuition, of attunement. If we want to speak of reason in this context, it cannot be instrumental reason, but a form of rationality which can grasp intrinsic value. It is not *Zweckrationalität*, but a kind of *Wertrationalität*, to use Weber's terms. Further, in a stance of disenchantment, we seek only *de facto* goods, things that are satisfying to our *de facto* desires. But what we are looking for . . . is our yearning for the intrinsic good. (270)

I do not want to suggest that Darwin's relation to nature was strictly Rousseauvian, for he had no doubts that nature is full of horrendous, even nauseating, phenomena, and, as we shall see, he has some stern Victorian words for it. But that he found in it "intrinsic" value is unquestionable. Nor do I want to suggest that adopting Taylor's version of the Romantic relation to nature fully re-enchants the world. The point is certainly not to ignore or minimize the alienation and sense of loss that have accompanied so much of modern experience, but to suggest that the experience of enchantment inheres within modernity. Instrumentalism is as central to modern experience as any exploration of its "intrinsic" worth; in all too many respects, the experience of enchantment I am discussing here is available primarily to those who have sufficient time and money in their private lives to connect with nature in noninstrumental ways. As in the development of Romanticism itself, with its vogues of "view-hunting," and its tourist's-eye view of the Alps, for example, a modern, naturalist re-enchantment threatens to be an evasion rather than a reengagment with modern experience.

A lot of bad stuff happens out there, quite naturally; some of the ideology to support that bad stuff derives from interpretations of Darwin. To imitate nature in its ruthlessness becomes a moral injunction, but to do so is of course profoundly immoral. As John Stuart Mill put it famously, "In sober truth, nearly all the things which men are hanged or imprisoned for doing to one another are nature's everyday performances."[40] The argument here is not an invitation to "follow nature" in any of the ways Mill exposes as meaningless or mad, and for which Darwin himself had to apologize at the end of the chapter entitled "Struggle for Existence." Nor does it invite an uncritical submission to nature's ways (whatever that may mean) and a silently awed reverence for its working. Our awe at nature needs to be tempered by knowledge of it and a recognition that human culture is usually in some ways at odds with nature, fending it off with air-conditioning and fire and television and computer technology (all of which, of course, are also using nature), making it easier, ideally, for at least one species to live in it. But in the face of the myth of disenchantment, which implies that meaning and value go out of the world as soon as it can be explained rationally and naturalistically, I want to support such explanation and at the same time assert that value inheres *in* the world so described, just because of our relation to it.

The excitement that follows upon understanding the instincts that drive birds to migrate (and this requires no mystification or invocation of transcendental spirit), the astonishment that follows upon recognizing the overwhelming complexity of the eye's functioning (even despite the flaws in the mechanism that are clear evidence that there is no intelligent design behind the construction of the eye), the recognition that living organisms are mutually dependent in ways that only the most delicate and careful investigation can discover—these and all the various knowledges that scientific study of nature and the human has been producing are elements of new forms of enchantment. The consequence of such knowledge is not—or should not be—the determination to do as nature does, whatever that might mean. Nor should it entail ignoring the dangers and deadening implications of mere instrumentalism. But surely the consequence is not to give up on the world because somehow it has lost its meaning—it is stunning, beautiful, scary, fascinating, dangerous, seductive, real. It offers itself as the occasion for enchantment that Weber and others have thought might only be experienced in a world imagined as teleological and transcendentally grounded.[41]

Darwin's relationship to nature may in this respect be taken as exemplary. He approached nature, yes, with Herschel's kind of Enlightenment intensity of rational curiosity and ambition to "explain," but with something else that only occasionally is attended to when the various "users" of Darwin set to work on their scientific or social programs. I want to suggest that Darwin's prose is extremely sensitive to the emotional effect both of what he is trying to argue and of the phenomena of nature with which he is continually engaged. That effect, or "affect," is not only reverential toward nature, but it emerges from a constant struggle with it to yield its secrets, and a detailed recognition of its perfidies (I use the term precisely because it seems so Victorian, as nature did to Darwin). To know, in Darwin's prose, is in a very important sense to feel. And no one more than Darwin (who couldn't stand the sight of blood and recognized natural horrors when he saw them) knew and felt the variety and beauty of nature and its almost infinite possibilities for growth, form, connection, and interaction. The disenchantment narrative is implicitly based on the assumption, so important in Weber's thinking, that fact and value are entirely distinct, that facts do not entail moral action.

The philosophical maxim that "is" cannot translate into "ought" is certainly an important one, and the failure to attend to it has led to some of the grossest misuses of Darwin. But on the other hand, in most practical circumstances, the division between fact and value is extremely artificial. Hilary Putnam has recently reconsidered the fact/value dichotomy, arguing that there is no "notion of *fact* that contrasts neatly and absolutely with the notion of 'value' supposedly invoked in talk of the nature of all 'value judgments.' "[42] "Value and normativity," he says, registering the views of the pragmatists, with whom, on this issue, he largely agrees, "permeate *all* of experience" (30). It permeates without apology the language of Darwin. If, as Putnam says, *"theory selection always presupposes values"* (31), it is not unreasonable to suppose, as Desmond and Moore have shown with historical evidence, that Darwin's "selection" of a theory was infused with values (although of course the nature of his "theory" is different from the sort of theory Putnam is discussing). But so too is his selection of details from nature to support his theory.

What I am proposing here in suggesting an "enchanting" use of Darwin is that while he too aspired to objectivity, his language entangles fact and value from the very start, not merely in the epistemological sense that it entails the selection of a theory to justify them, as all "scientific" arguments must, but in the sense that what Darwin looks at strikes him, as that language consistently reveals, as valuable and usually as morally loaded. My project in this book, then, is to develop a kind of heuristic for further explorations of the human satisfactions that Darwin's kind of materialism and secularism might produce in a world that on the one hand seems to have bought the narrative of disenchantment and, on the other, seems, yet more dangerously, to have gone slightly mad in its quest for transcendental consolation.

I will, then, take the risk of being sentimental about Darwin, his ideas, and his potentiality for cultural good. The sentimentality will, I hope, be offered with more than a little tough-mindedness, that is, without losing sight of Mill's recognition of the horrors that nature perpetrates, or of the kinds of critique implicit in Desmond and Moore's understanding of how the theory was developed, or of the understandable distrust of aesthetic satisfactions that has come to characterize much theorizing on the left: "any expression of attachment to the world," says Connolly, "is . . . chastized by being treated as incompatible with a commitment to social justice" (10). That chastising is what I, like Connolly, want to reject. Although this book will focus almost entirely on Darwin and his readers, underlying its arguments is the reverse assumption: that in fact, the injunction to be tough-minded and see the world with the cynical skepticism that it deserves is inadequate to motivate action for social justice. Seeing value where we have missed it, *feeling* value where we have not felt it, is a condition of pursuing value. Readings of Darwin that ignore the affective elements of his writing—and for the most part, these are the readings we are getting[43]—leave us with just the sort of disenchanted world that Weber describes and that, for example, Pinker's prose reinforces. But it is impossible to read Darwin without recognizing in him the deepest possible "attachment to the world."

Part of the very moving story of late Victorian literature and culture is the persistent effort by many of the most important intellectuals to come to terms with what was felt to be the bleakness of a world that apparently offered no compensation for the pains it always inflicted. Absolute secularism was a hard pill to swallow, and many strong secularist movements sought some of the spiritual solace that religion had previously seemed to provide. It wasn't easy for Darwin, either, as we have already seen. But the tendency to understand Darwin's world as providing no affective or even rational compensation is, from the point of view of this book, another of the "misuses"—although perhaps an inevitable one—of Darwin. Darwin's business was not consolation, of course, and yet in rewriting the Western myth of origins, translating it into evolution by natural selection, his writing attends with loving care to the particulars of organic life and bespeaks a profound passion for the world and its minutest denizens. Darwin's religion was in nature. His son William wrote that his "deep sense of the power of nature may be called in his case a religious feeling . . . he had no religious sentiment."[44] The texture of this feeling, deeply secular and intense, reveals that Darwin's work of sweeping away the teleology of natural theology and subjecting all biological phenomena to scientific explanation was nevertheless fully compatible with a sense of a world deeply infused with value, enchanted.

In this argument, I align myself with thinkers like Connolly and Jane Bennett, who have tried to rethink secularism beyond the Enlightenment tradition of pure rationality that has for centuries now been its intellectual armory. It was clear to the Victorians themselves that Enlightenment rationalism was not adequate to the real human needs of a culture from which the supports of traditional religious beliefs were being driven. Among them, the experience of religious loss evoked a wide range of responses, the best known being positivism itself. And yet Peter Allan Dale was certainly right when he claimed that positivism was perhaps the most important Victorian manifestation of the Romantic quest for "an adequate replacement for the lost Christian totality."[45] To be sure, even at its height, the "Positivist Society" could count among its London members only ninety-three people. George Eliot, though she was at times an enthusiast for Comte, and composed the famous Positivist hymn, "O May I Join the Choir Invisible," would not attend the Positivist Church. T. R. Wright reports the contemporary joke about the schism within the Positivist Society, that "they had come to Church in one cab and left in two."[46] The Comte whom Mill ultimately came to criticize and from whom Lewes partly withdrew quite literally transformed his system into a church and projected an authoritarian political structure, more or less on the model of medieval Catholic theocracy, and antithetical to Mill's liberal, democratic beliefs. But the story of the Positivist Church is perhaps the most forceful challenge to arguments for the possibility of a "nontheistic enchantment," that is, the attempt to imagine a fully secular enchantment.

Victorian Positivism was rationality gone berserk, one might say. But, so the narrative of disenchantment goes, only something like a berserk rationality could transform the hard news of a world gone secular into something inspiriting. "Who," asks Max Weber, "who—aside from certain big children who are indeed found in the natural sciences—still believes that the findings of astronomy, biology, physics, or chemistry could teach us anything about the *meaning* of the world?"[47] Here is Weber, as he builds his narrative of a disenchanted world, insisting typically on the fact/value dichotomy that Putnam attempts to explode: there is and must be, in the classroom and in practice, a radical division between scientific thought and political, ethical, and aesthetic value. Weber argues that science, precisely as it asserts that "there are no mysterious incalculable forces that come into play," expels "meaning" from the world. "One need no longer have recourse to magical means in order to master or implore the spirits, as did the savage, for whom such mysterious powers existed." This, Weber says, "means that the world is disenchanted" (5).

Turning to the nature that science reveals for "meaning" of the sort that matters to human life and its conduct is, for Weber, a chimerical enterprise. The alternative to disenchantment is not to be found in the world in which rationality is a determining value but in religion, an option impossible for secularists and for a sophisticated modernity. Weber doubts that "religious interpretations" add to the dignity of moral acts. "The fate of our time is characterized by rationalization and intellectualization and, above all, by the 'disenchantment of the world'" (11). If, then, we accept the terms of Weber's argument, we must see my effort in this book as the work of a "big child," for indeed I am trying to suggest that Darwin's writing, read with literary attention, can facilitate a form of nontheistic enchantment, without having "recourse to magical means in order to master or implore the spirits."

But there is a significant history, leading up to contemporary efforts like Putnam's, to locate meaning and value in the world of facts, early positivism—not logical positivism—having been one of them. Positivism failed in its aspiration to rational enchantment in part because it didn't deal adequately with the oxymoronic quality of its effort, and the later positivism of the Vienna Circle simply set reason and feeling so radically apart that a Weberian disenchantment was the only possible resolution. In response to the spirit of triumphant rationality that he found in the writings of the exuberant secularist W. K. Clifford, William James insisted that human beings need something more than this thin gruel, and that the "rational" is not the supreme value for most of us. The rational, he argued, is always bound up with human need and desire. He takes up Clifford's dictum (he calls Clifford "that delicious *enfant terrible*") that "It is wrong always, everywhere, and for every one, to believe anything upon insufficient evidence." James insists on the human complication of belief, the inevitable admixture of "will," of "such factors of belief as fear and hope, prejudice and passion, imitation and partisanship, the circumpressure of our caste and set."[48] For James, in the end, this is not a failure of human intelligence but a condition of it, and rational choice is always involved with desire and need; it makes no sense without it. The split between the intellectual and the affective, which is central to Weber's thesis, does not operate for James, since all intellection is involved in the whole person thinking. "There is in the living act of perception always something that glimmers and twinkles and will not be caught, and for which reflection comes too late" (*Varieties*, 497).

A large part of what we confront each day is not decidable on rational grounds alone, and when people choose among rationally undecidable options, it simply doesn't make sense to choose the bleaker one. "The thesis I defend," James says, "is, briefly stated, this: *Our passional nature not only lawfully may, but must, decide an option between propositions, whenever it is a genuine option that cannot by its nature be decided on intellectual grounds*" (*Will to Believe*, 11). And for all but a very few propositions, purely intellectual grounds won't do. James's concern with the texture of human feeling and need in the heart of the deepest philosophical and scientific questions was his nonpositivist response to the disenchantment of modern naturalism and secularism. On this account of the work of "thinking," which, as I have noted, has been updated by Connolly and his talk of the "amygdala," Weber's disenchantment is simply a misdescription.

Connolly pursues the question in James's vein, challenging

the secularism that Weber so confidently and bleakly describes. One of the quandaries of secularism, he claims, "is that its forgetting or depreciation of an entire register of thought-imbued intensities in which we participate requires it to misrecognize itself and encourages it to advance dismissive interpretations of any culture or ethical practice that engages the visceral register of being actively" (29). There is here a Jamesian revulsion from the rigorous intellectual priorities of a W. K. Clifford,[49] and Connolly's critique of secularism entails the broad recognition that "argument, rationality, language or conscious thought" are "always accompanied and informed to variable degrees by visceral intensities of thinking, prejudgment, and sensibility not eliminable *as such* from public life" (36). The "scientific" claim that one is free from these things is a dangerous invitation to a disguised authoritarianism, exactly the reverse of what an Enlightenment secularist would ostensibly want.

The point for my argument is that for Connolly and James there are, in the midst of the world from which the transcendent has been expelled, "little spaces of enchantment" (17). Weber does not describe such moments or their possibility, and his account of the disenchantment of modernity allows him to commit himself to precisely the sort of secularism about which Connolly complains. Certainly, he describes a real cultural and attitudinal change in modern Western culture, but he takes for granted the idea that enchantment is, as Jane Bennett points out, dependent on a teleological view of the world and a "divine creator" (12).[50] For Bennett, it is not only the natural world that provides those moments of enchantment that give the world value; nor does she believe that "enchantment" can be a permanent and total condition. It is rather a "peculiar kind of mood" (34). "I pursue a life with moments of enchantment," she says, "rather than an enchanted way of life" (12). It might be appropriate to call the moods of enchantment "spots of time," moments that, while they can be relatively rare in one's life, fill it with meaning and value, and evoke memories and connections that themselves, in the Wordsworthian tradition, become richer and fuller. The argument for enchantment then is not a reassertion of the enchanted world, Taylor's world of the "sacred," which Weber regards as lost; it is not a nostalgic claim that a culture dominated not by the claims of rationality but by premodern ideals of community and coherence and religious significance is somehow happier, more fulfilled and fulfilling, resident in a state of permanent enchantment. This is no place to debate the reality of that ideal past. Nor do I want to argue (neither do Bennett or Connolly) that those spots of time, those little spaces of enchantment, return one to the organic and pervasive enchantment supported by traditions and beliefs no longer possible to modern people. My argument is only that the world deprived of medieval ideals can be rich with value and lovable, at least partially, at least sometimes.

Bennett's description of what it is to be enchanted is worth attending to since I am claiming that Darwin's writing and experience, and our possible experience of Darwin and, through him, of the natural world, open up possibilities of enchantment. "To be enchanted is to be struck and shaken by the extraordinary that lives amid the familiar and the everyday" (7). "The mood I'm calling enchanted," she says, "involves . . . a surprising encounter" that contains "the pleasurable feeling of being charmed by the novel and as yet unprocessed encounter," and a "feeling of being disrupted or torn out of one's default sensory-psychic-intellectual disposition." Its effect is "a mood of fullness, plenitude, or liveliness, a sense of having had one's nerves or circulation or concentration powers tuned up or recharged—a shot in the arm, a fleeting return to childlike excitement about life" (5). There we are, if unapologetically, back to Weber's "big children."

I cite Bennett at length here because hers is among the strongest of the efforts, as yet rare, to make a strong case for the possibility of valuing the world without faith in "transcendental design, teleology, or a divine creator." Bennett wants to "erode the belief that an undesigned universe calls above all for a cold-eyed instrumentalism" (34). Searching for value *in* the natural world need not take us to the bleak and cruel vision that Mill describes in "Nature," or to the merely formulaic, heartless world in which morality depends on the "prisoner's dilemma"— bound as we are by the ethics of natural selection. All altruism need not be reciprocal; real altruism, as many of even the sternest sociobiologists concede, is possible to us.[51] Following her claim that "you have to love life before you can care about anything" (4), Bennett argues that the "cultivation of an eye for the wonderful becomes something of an academic duty" if, as is often argued, it "can foster a laudable generosity of spirit" (10). It is not merely an aesthetically self-indulgent condition she is trying to describe. Bennett, like Connolly, is a political philosopher, and her concern with enchantment follows upon Weber's construction of the narrative of disenchantment and is aimed not at some solipsistic aestheticism, a museumlike experience of the wonders of nature or of technology, but at the cultivation of social generosity. The narrative of disenchantment, she claims, is not merely "a story," but an act in the world. It has consequences.

From my point of view, its major consequences are two. First, it leads to a consistent undervaluing of contemporary experience, a sense that the new—product of technology, science, social planning, and the rest—constantly drifts away from the great traditional values that gave meaning to life in premodern times, and a sense that nature itself has been drained of significance. (It also implies a "golden age" view of history, a deep nostalgia for a past that might never have been as "golden" as retrospect and contemporary frustrations imply.) Second, it assumes that all meaning and value derive from religion, and from a teleological view of the world. Without them the world is doomed to a breakdown of community against the forces of instrumentalism, rapacity, social confusion, globalization and homogenization: an ultimate Weberian meaninglessness. These attitudes are in a sense self-confirming, for acting as though there is nothing valuable out there but the instrumental is likely to foster the dog-eat-dog sort of world that has so often been called "Darwinian."

Although Darwin too was uneasy enough about the possible implications of his theory that he opened the door wide to the idea of progress (at least for the life of humanity), his argument in *On The Origin of Species* seems to preclude it. There is a chilling paragraph near the end of *Descent* that suggests something of the imperfection of the world Darwin imagines, of the possibilities for perfection that it allows, and the deep human significance that inheres in it and his relation to it.

> As natural selection acts by competition, it adapts the inhabitants of each country only in relation to the degree of perfection of their associates; so that we need feel no surprise at the inhabitants of any one country, although on the ordinary view supposed to have been specially created and adapted for that country, being beaten and supplanted by the naturalised productions from another land. Nor ought we to marvel if all the contrivances in nature be not, as far as we can judge, absolutely perfect; and if some of them be abhorrent to our idea of fitness. We need not marvel at the sting of the bee causing the bee's own death; at drones being produced in such vast numbers for one single act, and being then slaughtered by their sterile sisters; at the astonishing waste of pollen by our fir-trees; at the instinctive hatred of the queen bee for her own fertile daughters; at ichneumonidae feeding within the live bodies of caterpillars; and at other such cases. The wonder indeed is, on the theory of natural selection, that more cases of the want of absolute perfection have not been observed. (472)

There is perhaps nothing in the passage that closes out finally the possibility of "perfection" and progress, a kind of teleology, for the "naturalised productions of other lands" might constantly press toward an increasingly perfect adaptability that would ultimately make further invasions "from other lands" unsuccessful. And yet that is a story Darwin does not tell, and this initial statement seems implicitly to disown these possibilities. Adaptation, which in Paley always implied perfection, is "only in relation to the degree of perfection of their associates." Changes in the "associates," changes in the climate, detour evolutionary development, cut off some species, start producing others. Teleology here seems a very remote conception. And "absolute perfection," which the passage implicitly suggests is possible and even common—a concession to a Paleyan vision?—

becomes self-contradictory, since all species are "perfect" only to the degree that they occupy an entirely stable and unchanging habitat. That Darwin understood the fragility of this "perfection" is clear from the self-protective way he avoids, in the *Origin*, talking about the most perfect species, humans.

But more striking than the explicit argument is the texture of the catalogue of horrors and monstrosities the rest of the paragraph provides. For while on the one hand, this is hardly an "enchanting" litany, it has a characteristic Darwinian resonance of the sort this book will be most interested in detecting and discussing. In the first place, it is important to register the degree to which the paragraph depends upon and appeals to feeling and to fundamental human relations and expectations. The framing sense of the list to come is "surprise," which, to be sure, Darwin is ostensibly intent here on discouraging: "we need feel no surprise." But we do feel it, Darwin knows it, and he feels it too. The Paleyan reader, who holds the "ordinary view," won't know how to cope with the list. "We ought not to marvel," says Darwin, who is almost always marveling, and he then proceeds to list a series of "marvels": the self-destructive bee, "the astonishing waste," the dreadful parasites. The language is almost aggressively anthropomorphic, as if to emphasize both the fact of consanguinity with the human and the moral horror that marks so much of the natural world: the drones are "slaughtered," "the queen bee feels instinctive hatred for her fertile daughters."

It is not a pleasant vision, but the fact that it is registered as so powerfully moralized is significant. If, on the one hand, such a passage can encourage the notion of a natural world that lives out the "red in tooth and claw" vision so common to readings of Darwin, on the other, it gives us a sense of the world as thick with value. It is not empty and meaningless, but startling, frightening, entangled in ethical value and ethical struggle. All of these things, moreover, are wonders, as they extend our sense of the possible, and shock us.

The list Darwin offers is, he notes, "abhorrent to our ideas of fitness." As virtually everywhere in his work, Darwin takes for

granted the importance and even the inevitability of what he regards as his culture's assumptions. His approach to the natural
world is entirely from the point of view of a Victorian gentleman,
and while this has often been taken as a mark of the ideological
complicity of his science with the dominant conservative powers
of Victorian society, I want to suggest that this way of seeing also
opens possibilities for fresh and creative thinking, infuses the
world with value, implies the ethical significance of natural phenomena, and leaves the world fundamentally enchanted. Darwin
does not scrupulously depersonalize his writing, although he
certainly strains to give to his arguments and descriptions an objective substance; he writes like a scientist *and* like a caring, loving, conventional, and reverent man whose relation to nature is
intense and charged with feeling. Fact and value hang together in
the rhetoric and in the scientific imagination.

He does so even as he increasingly moves away from the religion that he had more or less conventionally accepted in his
youth. This chilling list of nature's monstrosities suggests in its
detail how his deep valuing of life in all of its complexity did not
entail a radical sentimentalizing, but rather incorporated into itself the full possibility for nastiness in nature that Mill was to
describe as criminal. In the parts of his *Autobiography* that were
originally omitted by Emma Darwin, Darwin talks about religion. There he argues that happiness and enjoyment of life are
not incompatible with "belief in natural selection." In fact, natural selection helps explain, as religion never satisfactorily could,
the suffering in the world that so disturbed Darwin. It is not, of
course, a justification, but it helps one understand—it *means.*
Natural selection "is not perfect in its action, but tends only to
render each species as successful as possible in the battle for life
with other species, in wonderfully complex and changing circumstances."[52] Imperfection and wonderfully changing circumstances are the conditions of Darwin's world.

While there is perhaps some little casuistry in Darwin's attempt
to argue that natural selection guarantees the predominance of
happiness over suffering, he surely is straightforward in his

view that natural selection accounts for why the world is so full of suffering and horrifying activities, like those of the ichneumonidae. In the intensity of his engagement with the natural world, Darwin offered to his readers one of the very richest possible compensations for the imperfections, cruelties, and indifferences that his studies seemed so often to reveal. Reading his work with care, one will find, as Robert Richards has recently argued, that far from proposing a world that mechanistically functions without spirit or moral compass, Darwin's writing belongs to a great tradition of romantic literature and thinking that imagines nature, with all its obvious horrors, as essentially benevolent and altruistic—quite the reverse of what many modern uses of "natural selection" describe.[53] The point for me is not to urge assent to the vision of this kinder, gentler Darwin but to understand how his writing becomes an excellent model for the collapse of the fact/value dichotomy and a major indication that caring about the world, feeling its powers of enchantment, is fully compatible with a scientific approach that refuses to move beyond naturalistic explanation.

Darwin is not kidding when he claims that "When I view all beings not as special creations, but as the lineal descendants of some few beings which lived long before the first bed of the Silurian system was deposited, they seem to me to become ennobled" (*Origin*, 489). Nowhere in Darwin does one find the revulsion from things in this world that makes worldliness seem immoral. He never felt disturbed that his ancestors were the "lower" animals; the imagination of that development was, instead, thrilling. Darwin's world, while it points always toward that naturalistic explanation, pushes frequently also toward the sublime, toward that dizzying vision of endless time, of staggering complexity, of interdependence and paradox, that replaces the "enchantment" that a divinely constructed nature has been said to produce.

Jane Bennett's project of insisting on the possibility of spaces of enchantment in this "disenchanted" world does not confine itself exclusively to nature. While Darwin's prose is obviously not related to modern technology, it might be useful (because, I think, it is in the Darwinian mode) to consider her arguments about modern enchantment beyond "nature." Enchantment for Bennett is particularly an effect of the sort of metamorphoses that historians and philosophers like Donna Haraway and Bruno Latour describe. "Late modern morphings and Paracelsian interminglings are uneasy admixtures of organic, fantastic, commercial, scientific, and moralizing forces. By drawing parallels between these two sets, the enchantment effect of the contemporary morphings might be enhanced" (50). The world of disenchanted modernity, on Bennett's accounting, turns out to be a world of transformations.

Darwin's is a world of transformations. He does not, of course, offer us cyborgs, but after the *Origin* the culture had to confront the possibility that humans bore within them the genealogy of nonhuman beasts and that they, like all other living creatures, were potentially in a condition of transformation. Early in *The Descent of Man* Darwin inserts a pair of illustrations. The first shows the embryo of a human, just above the embryo of a dog. Darwin counts on the shock of juxtaposition, the startling similarity between the two. It is as though the one "morphs" into the other, and Darwin quotes Huxley: "the mode of origin and the early stages of the development of man are identical with those of the animals immediately below him in the scale."[54] That Darwin insisted on human descent from "lower" organisms is now a familiar point, of course, but it needs to be set here in juxtaposition with Bennett's argument, which locates moments of possible enchantment in the constantly morphing conditions of modernity.

The second illustration has a similar, if somewhat subtler, psychological effect, but its power of fascination remains as strong now as in Darwin's day. It is an illustration of a human ear with an arrow pointing to "a little blunt point, projecting from the inwardly folded margin, or helix." Darwin shows that other anthropoids have similar points, and since, as he claims, "every character, however slight, must be the result of some definite cause," he regards the "point" as "a vestige of formerly pointed ears" (23). Everyone I know who has read the passage and all my friends on whom I have tried it out reach immediately for their ears, rub the point, and feel their connection with pointy-eared ancestors. It is the felt character of this movement that I find so striking. If the feeling is not enchantment, I am not sure what to call it, that instant of surprise, and the extraordinary visceral sense that one is somehow in contact with a past that reaches back millions of years and at the same time connects one with all other living humans and all other mammals, too. But the moment comes to us in the cool, objectivist language with which Darwin develops his arguments, and is methodologically connected with other moments when anthropomorphism and unrestrained affect enter overtly. "Must be the result of some definite cause" in this sequence suddenly explodes into the visceral connection with the deep past.

Darwin's world of change and crossings, where essential categories are constantly disrupted, is a world that includes those spaces for enchantment that Bennett discusses. And in such a world, matter, so often set up in a dichotomy with spirit, comes alive. "The problem of meaninglessness," Bennett says, responding to Weber's notion that science drains meaning from the world, "arises only if 'matter' is conceived as inert, only as long as science deploys a materialism whose physics is basically Newtonian." But in Darwin's language, as in Bennett's narrative of re-enchantment, "matter has a liveliness, resilience, unpredictability, or recalcitrance that is itself a source of wonder for us" (64). If Darwin is taken as the kind of patron saint of a dog-eat-dog, ethically meaningless world, he needs to be seen as well as the patron saint of a world where matter is in constant motion, constantly transforming, constantly producing variations and surprises, manifesting stunning connections. The transformations that I see as deriving from a Darwinian understanding of time, organism, and change are for Bennett precisely the sort that open up modern experience to enchantment. In her discussion of natural phenomena—plants and aphids and ladybugs

(170), for example—she focuses on interrelations and transformations that closely resemble those that Darwin studied, although Darwin does not enter into her discussion. "Nature enchants," she says, "but so do artifacts."

Darwin does not deal with the artifacts, nor will I. But he deals with contrivances and morphings and disturbances of the ordinary, and, as we know, "there is grandeur in [t]his view of things." Darwin demonstrates as richly as any writer in the language the ways in which affect and intellect, value and fact, are aspects of the same phenomena. His natural world breaks down the absolute borders that separate species from each other, puts the world in motion, opens sublime vistas of past and future, ennobles a humanity that constantly threatens to denigrate the body (a product of millions of years of complex development) and submit itself to some noncorporeal Other beyond the reach of time and change. Darwin's world—which is our world—is an enchanted one, if we would allow ourselves to look with his eye for detail and aberration, movement and connection, and his reverence for living things; if we could detect the serpent in the bird, the woman in the man or the man in the woman, the caterpillar in the moth; if we could learn to read in the facts of the moment traces of the past and intimations of a future; and if we acquired the strength to confront without metaphysical equipment the astonishing richness of the material world. Darwin loved his dog, his pigeons, his garden, his family, and the world, and we are all part of that world.

CHAPTER 2

The Disenchanting Darwin

If I am, in this book, to return to the idea of a redeeming Darwin, of a Darwin whose cultural power might be taken as both humane and enriching, it will be necessary, as I want to do briefly in this chapter, to face this culturally saturated Darwin precisely in the places that implicate him in his culture's prejudices, and that have issued out in various social and political movements that seem to have had very unhappy consequences. Since I will want to be arguing that Desmond and Moore are right, that Darwin was indeed very much a man of his moment, but that being a man of his moment was a positive condition of his best thought (as well as of his worst), I will have to look at the kinds of views that convinced some Darwinians to ignore or minimize them and that partly justify some of the most unfortunate uses of his ideas by others.

Darwin's work is marked by two qualities that might seem particularly disenchanting: a constant impetus toward transforming mysteries—particularly about the human condition—into "problems," and then pushing forward to at least tentative solutions; and toward explanation of development in the natural world in terms that seem to be translatable immediately into the kind of politics that, for example, Ridley overtly adopts.

The recent historiography to which I have already alluded has made it more difficult to ignore Darwin's participation in his culture's prejudices, his hierarchical sense of race, his belief in the superiority of his own class, his view that women were intellectually inferior to men. Robert Young was angrily outspoken on the subject, complaining, long before the appearance of Desmond and Moore's biography, that scientists and historians had carefully separated Darwin from the historical context in which he was "enmeshed in a tight web of social, cultural,

and ideological determinations."[1] The historical Darwin is
quite another man from the more purified, almost hagiographi-
cally treated genius whose theories are seen as growing only
from the internal logic of his discipline. His motives are not so
pure, his thinking not exclusively related to strictly "scientific"
concerns (given the possibility that we might fully detach scientific from other concerns), and he was not free of blame for
some of the uses to which his ideas have been put—from free-
wheeling capitalism to eugenics. As Desmond and Moore show
convincingly, his ideas were intimately connected with the
development of laissez-faire capitalism in the nineteenth century. "The Darwin-Malthus connection is now firmly established," wrote Young in 1987, and the idea of natural selection
has served the purposes of everyone from Walter Bagehot to
Adolf Hitler.

But I want to insist that the social analysis is not *all*, that the
ideas are anchored in their own rational and empirical justifica-
tions, and that insofar as they are, they are not tied *intrinsically*
(I repeat and emphasize this word) to the cultural prejudices
that informed them. Darwin's moment, his prejudices, his reactions to the dominant ideas of his day all played a critical part in
the formation of his theory—without them, in fact, that theory
would surely have a different shape. But they are not intrinsic to
the theory in the sense that one cannot hold the basic ideas we
might call "scientific" without also holding the prejudices and
cultural assumptions out of which they emerged.

So saturated in cultural value are Darwin's ideas that it is often hard for cultural critics to take into account the most obvious
fact about them, how enormously valuable they have been as
"science." The trickiest aspect of Darwin studies is somehow to
avoid, on the one hand, the tendency to separate out the scientific from the cultural in his thought, as though they were absolutely distinguishable all the way down, and, on the other, the
tendency to act as though a sociological explanation might account *fully* for the nature of his ideas.[2] Part of the great challenge
in studying Darwin is somehow to reconcile two ostensibly con-

tradictory epistemological principles: the first, that one cannot move from description to prescription, from "is" to "ought,"[3] and the second, that every assertion of fact is laden with prior theoretical and cultural assumptions.

At the same time, however, while virtually any of the many uses of Darwin can be seen to have at least some ground in what he actually said, some uses are closer to his original meaning than others. Some uses are very close to *a part* of what he said but manage to ignore or explain away other aspects of his thought. It is always hard, for example, for current scientific Darwinians to face the fact that Darwin included the inheritance of acquired characteristics in his explanation of the origin of species, and increasingly as he revised the *Origin* later in his career.[4] But one of the central elements of debate (on which both sides are partly right) is whether his theory entailed a constant war among contending species, contending individuals, contending males, or whether it argued for the importance of mutual dependence and cooperation. Altruism has become one of the key problems in modern evolutionary biology, and the debate often circles around the argument about whether evolution works through individual or group selection.[5] Cultural issues here depend upon the scientific ones. Whatever we think about the is/ought dichotomy, in the history of post-Darwinian thought, "is" and "ought" are almost always intertwined.

But it is necessary to insist on the presence in Darwin's thought of what I will now call "cultural" forces, always in quotation marks, to indicate the inadequacy of the "science/culture dichotomy," and to allow not only that Darwin made mistakes (at least from the point of view of contemporary science)—hardly a dramatic claim—but that much of the nasty business that has borrowed the name "Darwinian" has genuine roots in what he wrote. On the road to enchantment, it's critical that one recognize that the Darwin red in tooth and claw, the Darwin of a kind of outlaw capitalism, the Darwin as father of eugenics is not in any way an inevitable Darwin—except that any theory of such broad and world-historical significance is likely to be taken in

dangerous directions, always depending on the particular historical circumstances in which the ideas are received.

I

Recently, as I was reflecting on these problems and about the
way the word "Darwinian" is used these days, I was delighted
to find on the op-ed page of the *New York Times* the headline
"Let's Leave Darwin Out of It." The essay was by the late ubiquitous Stephen Jay Gould and was aimed at only one aspect
of the ubiquitous Darwinizing. Gould there attacks what he calls
"the current fad in conservative intellectual circles for invoking
the primary icon of my professional world," Darwin, as "either a
scourge or an ally in support of cherished doctrines."[6] Since it is
hard enough to get Darwin right, and since Darwin himself was
not always consistent, Gould's effort to get him out from under
the pressures of contemporary conservatism are admirable, and
I would in many respects like to have joined him in that effort.
But surely, of all people, Gould was aware that this is not a new
phenomenon. His attempt to disentangle Darwin from political
uses feels just a bit disingenuous. Surely Gould had read the
Robert Young who proclaimed long ago that "Darwinism *is* social,"[7] and the Desmond/Moore "full blooded" biography of
Darwin, and he knew of the tradition of Darwin studies that
these writers may be said to represent, a tradition that makes it
very difficult indeed to think about Darwin's ideas without recognizing the social and political engagements and evasions they
embody. In a recent response to critiques of their biography,
Moore and Desmond deprecate what they call the "impressive"
attempts to "decontaminate natural selection."[8] They believe they
have the smoking gun, and their arguments are powerful—as a
concept, "natural selection" is deeply embedded in the social
and political constraints of Darwin's moment.

   Yet Gould writes as though Darwin's "evolution" had not immediately fed political interests at the moment of its publication,

or as though, to take just one obvious example, Walter Bagehot's *Physics and Politics* hadn't in 1872 used Darwinian theory as a defense of liberal democracy and the sustaining of social order. Bagehot is explicit about the connection, arguing that "as every great scientific conception tends to advance its boundaries and to be of use in solving problems not thought of when it was started, so here [with natural selection], what was put forward for mere animal history may, with a change of form, but an identical essence, be applied to human history."[9] As Alvar Ellegård argued, the broader public's reception of Darwin depended on the "religious and ideological implications of the theory."[10]

Thus, while one might want to rip Darwin out of the hands of conservative politicos, it would be a mistake not to recognize that to a certain extent he has always been in those hands and that there has always been something about the Darwinian program that has invited political application, even by Darwin himself. Darwin went on record connecting human to animal biology *and* behavior. However scientific the project, there can be none more latent with ideological possibilities. Darwin's philosophical positioning, claims Greta Jones, arose from his direct "confrontation with the question of human development and social evolution."[11] How might one do that without getting involved in social and political questions?

Gould's left-of-center efforts to separate Darwin from the idea that, as he quotes John O. McGinnis, "a Darwinian politics is a largely conservative politics" are in this respect similar to right-of-center efforts, which have historically been much more common, to enlist Darwin in the support of right-of-center projects.[12] Gould—sustaining the "is" can't be "ought" position—insists that science can never decide "the morality of morals." But clearly, *The Descent of Man*—the most obvious example among Darwin's works—at moments gets very close to suggesting a social program, makes moral judgments, projects a social future, and runs throughout with the cultural prejudices of Darwin's moment. It is not merely that Darwin makes some "scientifically" disputable arguments, as for instance that men are intellectually superior to women as clearly as they are physically stronger. It is that he is ready to take his facts—though certainly not in all of his books—and apply them to social conditions in ways that imply or suggest social action. Not surprisingly, these implications come close to explicitness only where Darwin dares address the subject of human development directly. To be fair, he is cautious and responsible as he edges toward areas in which his empirical evidence is incomplete, so that, when he argues that similar education would probably not bring women up to the level of men because of the disparity in their natural powers, he pauses parenthetically to admit, "I am here wandering beyond my proper bounds" (*Descent*, 2:329). He knows that the experiment has not yet been tried.

Darwin's naturalization of what most people think of as the distinctive characteristics of the human species—aesthetics and morality—however dispassionately articulated, was certain to provoke, as it did, quite intense reaction and lead to ideological positioning that continues even today. On this point, Alfred Russel Wallace separated himself from Darwin and kept at the center of his project a moral and political objective. From the very start, as Martin Fichman shows, Wallace was committed to a "scientific theism," and that in turn was compatible with Wallace's growing socialism.[13] But Darwin, even at the outset of his speculations about the distinctively human, about ethics and art, had been doing empirical work, carefully noting the progress of his firstborn child, for example, filling his notebook "M," begun as early as 1838, with ideas about the biological origins of behavior. As Sandra Herbert and Paul H. Barrett note, "Darwin's comments on the origin of man reveal that the subject held no terror for him."[14] And the "M" and "N" notebooks are full of entries that detect human qualities in spiders and puppies and monkeys and virtually any other sort of animal. If such animals are capable of communicating and expressing emotion, are they not likely to teach us something about ourselves as humans, to bear within them the instincts that have produced our emotions, our thoughts, our ideals?

In the long history of Darwinian thought, from Spencer to Huxley to E. O. Wilson, scientific naturalism has always also meant that biology is a form of social science and of ethics.[15] What may be the consequences of believing that ethics are built on biological imperatives? Is it possible to derive from those imperatives an ethics that resists or denies those imperatives? Against the sociobiology of his own time, T. H. Huxley in *Evolution and Ethics* urged, "That which lies before the human race is a constant struggle to maintain and improve, in opposition to the State of Nature, the State of Art of an organized polity."[16] In effect, Huxley was arguing that morality must resist biology—despite Darwin's view that it follows intrinsically from nature and natural selection, a point with which Huxley really agreed.

Huxley was struggling with the irony that the dominance of strictly biological explanation for all things human—determinedly neutral on ideological issues—slides the ideological into the biological; this leaves virtually no place to assert the moral against the biological, "culture" against science. Among the many different ideological applications of Darwin's theory there have always been some that insist on their neutrality even as they in effect give their support to intensely partisan political movements. At present, the strongest, and potentially most difficult version of this tradition is variously known as sociobiology or, as it has developed, one might say, "more scientifically," evolutionary psychology, to which, in this book, I will take many opportunities to return.

Many working in these areas describe themselves as "reductionist." Following upon the traditions that explored and justified scientific work in the nineteenth century and that issued out into various forms of positivism, evolutionary psychology, which is built on a strictly Darwinian program in which natural selection is the key, often commits itself to the heuristically valuable notion that science proceeds by reduction. All phenomena of human consciousness and behavior would then be understood to be explicable in terms of biological phenomena, and the biological phenomena might be reduced to the smallest possible unit, like the gene, or the DNA material that composes it, and the workings of natural selection. In a certain sense the reductionism is only an updated version of Darwin's own arguments that we can understand morality as growing out of "the social instincts," the herd behavior that served to minimize the danger to groups of animals through the tendency of individual members of the group to announce, even at their own risk, the presence of predators.

Darwin never talked about himself or his method as "reductionist," but his work has elements strongly compatible with reductionism. Early apologists for science, and Darwin himself, understood science to progress as it brought together empirical facts into increasingly generalizable laws—an escalating induction, as it were. The largest phenomena would then be explicable ultimately by the combinations of lesser and more basic ones. General laws with universal applicability might be derived from observation of and experiment on particular phenomena. Darwin's progress from the communication among spiders to human speech is potentially a reductionist argument.

There is much to be said for the usefulness of this kind of thinking, and it is obvious that Darwin's thought flourished within a framework that allowed him to imagine modern species as developing from simple, even unicellular life—although this of course is not exactly the language he used. Reductionism has done important work: it has led many scientists to see the interaction of phenomena, and it has built an increasing interdisciplinarity among sciences. This, certainly, has been one of the inheritances from Darwin, who, setting forth on the *Beagle* with a head full of Charles Lyell's geology, kept asking larger questions about the phenomena he noted, until he moved through geology to biology to anthropology and found a string of connections across virtually the whole range of knowledge.

Sociobiology today can be seen legitimately as one inheritor from Darwin. It imagines itself not merely as a bringing together of social and biological study but as a move in the direction of the ultimate unity of all knowledge.[17] All macrophenomena, from stars to morality to literature, might in the end be reducible to microphenomena universally spread through all of nature and culture, and culture would then itself have to be recognized not only as a part of nature but entirely explicable through those universal microphenomena. This is a powerful move that feeds the disenchantment narrative forcefully. Reductionism is, after all, just the procedure to remove meaning from the world, to exile the sacred in the quest for the entirely naturalistic explanation. Wilson prepares us for the bad news, insisting that rejecting reductionism is "the white flag of the secular intellectual" (*Consilience*, 297). We may not like it but we are, down to our finest habits and our highest arts, biologically determined (he would not quite use that term), and at the end of *Sociobiology* he invokes Camus, at once trying to reveal his own broad culture and invoking that culture to get us to accept the bad news: "A world that can be explained with bad reasons is a familiar world. But, on the other hand, in a universe divested of illusions and lights, man feels an alien, a stranger. His exile is without remedy since he is deprived of the memory of a lost home or the hope of a promised land" (301). Reductionism, Wilson suggests, will offer us no lost home but will give us knowledge of a past that will feel alien to us—there is no Eden at all.

Much current resistance to reductionism is based on the argument, with which this book will be recurrently concerned, that reductionism, in assuming that all complex structures are built up out of smaller and less complicated ones, tends to ignore the problem of the relations *among* the smaller structures; it fails, then, to address the problem of how the relationships among the basic phenomena actually change the way those phenomena behave. Richard Lewontin is one of the strongest critics of reductionism in this respect. DNA and genes, he claims, are inadequate to explain the full complex of human behavior *and* human biology, precisely because every such unit exists in a singular environment that affects the way the gene acts and how DNA messages are decoded.[18] Lewontin is certainly a Darwinian, but he is also certainly not a reductionist, and while it is clear that Darwin can be assimilated to a reductionist interpretation, it is also evident that reductionism is not an inevitable consequence of his arguments. It is possible to accept Darwin's evolutionary theory, as for example Loren Eiseley does, and resist the reductionist implication, which Eiseley regarded as a kind of mechanizing of life—a mechanization that, it is necessary to note, he found in Darwin's formulation of the theory itself.[19]

In the end, not only would human behavior be explicable by biology, but biology would be explicable by chemistry, and chemistry explicable by particle physics. Thus, Steven Weinberg claims to be a reductionist in his *Dreams of a Final Theory* because, as he asserts, that's the way nature is: "reductionism" is a "statement of the order of nature."[20] But even reductionism is not a simple thing: as John Dupré shows,[21] there are several kinds. At the center of any reductionist program, however, is the idea of a hierarchy in nature and thus a hierarchy of explanations. The largest, to put it crudely, will be explicable in terms of the smallest. Reductionism gives causal priority to one level of explanation over all others, so that, in the matter with which I am concerned here, macroorganisms become less real than microorganisms and can become causally irrelevant in explanation. The play of forces that Darwin saw at the macro level, among full-blown organisms, is now seen to be playing out at the very smallest level, at the level of genes and DNA, where it becomes hard to recognize human life and society as we know it at the normal, that is, the macro, level. At the lower level the difference between biological and cultural explanations disappears, and the biological impinges powerfully on the moral, because if at this level the biological is determining, it is simply immoral (or at best amoral) to argue, as Huxley did, for behavior counter to what biology demands. Dupré, resisting such reductionism, argues that "as objects are united into integrated wholes they acquire new properties," and he "sees no reason why these higher-level wholes should not have causal properties just as real as those of the lower-level wholes out of which they are constructed" (162–63).

But at whatever level of discourse about the human condition one moves, Darwin is there, and the epistemological battle about reductionism is also, obviously, a battle about social possibilities. Reductionism itself is not self-evidently a politically conservative epistemological theory, but as in Darwin's day, with his own ideas, the issues move from epistemology to questions of determinism, the alternative pressures of nature and culture, the limitations imposed by sexual difference, the possibilities of large-scale social change, and so on. The interpretation of Darwin that lies behind this reductionism, tending toward biological determinism, fairly enough points to his constant insistence on general laws as explaining even the phenomena for which he cannot find general laws. "The unknown laws of . . ." is a familiar enough phrase in Darwin's prose. He rejected the idea that chance operated anywhere in the universe, although, of course, much of the revulsion from his theory was a response to the major role that chance seemed to play in it. Darwin always seeks laws and was shocked when it was reported that John Herschel, after reading the *Origin,* called it the law of higgledy piggledy. "What this exactly means," wrote Darwin, "I do not know, but it is evidently contemptuous—If true this is great blow & discouragement" (*Correspondence,* 7:423). The universe Darwin sought to describe was law-bound, and it is possible to infer from his arguments what he surely never affirmed and, I believe, did not intend to imply, that biology fully determines all human behavior.

Sociobiological reductionists tend to insist on a determinate set of scientific meanings for Darwinian ideas and use the science as an impregnable authority for a biologistic rereading of human behavior, consciousness, and society. They and their descendants, evolutionary psychologists, deny that they are commited to the idea of biological determinism, but certainly they are committed to the view that the biological inheritance is so powerful that it significantly limits the possibilities of human freedom.[22] Biology, in this mode, quickly tends to become ideology, claiming all the while to be innocent of anything but stern and disinterested readings of natural fact. We have seen how

Wilson invokes Camus to make the point. We must take a good look at the worst. If their conclusions seem to have ideological implications, they can mean only that what has traditionally been taken to be ideology is simply the way things are: Huxleyan resistances become mere contradictions in terms, or symptoms of feeble thinking. Reductionists are above all insistent that their work is *not* ideological—that it is simply scientific.

Although the argument that sociobiology—or its derivatives—is not ideological seems to me either disingenuous or simply wrong,[23] the problems that these projects raise might be taken as reasonably derived consequences of Darwin's naturalism. Since many who are opposed to sociobiology and to "reductionism" are themselves committed to full-scale naturalism, there seems some inconsistency in resisting the notion that social, psychological, and behavioral activities might be understood in the terms of a hierarchically structured science. Wilson's lament that attacks on reductionism are the white flag of secular intellectuals accurately points to the fact that the wars against sociobiological theory have been conducted primarily by secular intellectuals who, like Darwin and like Wilson, also utterly reject metaphysical or nonnatural explanation. For Wilson, rejecting reductionism means rejecting science, because for him science means the idea that all things can be explained without metaphysics, by natural causes. But of course, there are many scientists, like Lewontin, who are not reductionists. Contemporary reductionism is replaying the kind of Victorian scientism that Huxley—perhaps scientism's most powerful voice—was most responsible for propagating but which he found himself resisting at the end of his long career.

One of the ironies of Darwin studies is that while interpreters rush to him because the scientific status of his theories gives them particular power in cultural arenas, the polysemic nature of his arguments and language makes a determination of what he really meant rather unscientific. Beyond the work of sociobiologists and evolutionary psychologists, whose rigorous commitment to the dominance of natural selection as explanatory of biological development is usually in reductionist form, the range of possible interpretations seems boundless, although many along that range seem to me just wrong (and seemed so to Darwin, too, as a matter of fact).[24]

Sociobiology and evolutionary psychology claim status as sciences, and it would be absurd to insist that they are merely ideological, the product of a particular "culture." But they continue to develop a single strain of Darwinian thought, which, though it has been present from the start, is fundamentally mechanistic, assumes the absolute primacy of empirical and rational choice, and moves if not toward biological determinism at least to a view that all the aspects of human behavior can be largely explained by biology. Obviously, not all sociobiologists or evolutionary psychologists can be seen in this way, but by and large, the scientific traditions they uphold have about them what Loren Eiseley has identified as the most repulsive elements of Darwinian thought: "its constant emphasis on struggle" and "its mechanistic, utilitarian philosophy which, to many, seemed as dingy as a Victorian factory" (*Darwin's Century*, 250).

But that very tradition has issued out in other ways. There remains the paradox: the view that his ideas are linked to the particularities of his moment, of his class, to the assumptions of his Whiggish politics, to his life in a wealthy Shrewsbury home, to his marriage with yet another Wedgwood, to his accommodations to the Church at Down, to his anxieties about class (Janet Browne shows, for example, that he was careful to be photographed without any implements of his craft, as a man without work, a gentleman);[25] and the view that his thought has revolutionary or at least liberating implications and has wedged its way into the very foundations of Western antiessentialist and antimetaphysical thinking. That is, Darwin has provided a continuing occasion for the dismantling of metaphysics, as is clear in the way he figures so prominently in the doggedly mechanistic and computational arguments of Wilson and current science of mind. At the same time, since arguments need some ground on which to build—even if it is turtles or elephants all the way

down, we have to stop at one of those elephants somewhere—
Darwin's theory has, ironically, become a new foundation. One
notes that even the rigorously and aggressively relativist Barbara Herrnstein Smith accounts for the "endurance" of literary
works through a Darwinian evolutionary model.[26]

As "natural selection" has become an indispensable, grounding concept, it is enlisted to do all sorts of (sometimes incompatible) work. It is, Pinker asserts, "the only evolutionary force that
acts like an engineer, 'designing' organs that accomplish improbable but adaptive outcomes" (*How the Mind Works*, 36, 155).
Arguing for a "computational theory of mind," Pinker regards
natural selection as "indispensable to understanding the human
mind." It is a "foundation," for it "alone explains what makes
life special," its *"adaptive complexity* or *complex design"* (155).
"Natural selection" becomes foundational because it replaces
the telos of traditional causal explanation. As Henry Plotkin describes it, the nature of all organic beings is determined by their
purpose, the reproduction of themselves. Darwinian explanation provides a posteriori rather than a priori explanations of the
adaptations that allow the organism to survive and reproduce itself,[27] and Plotkin is quite explicit about the way in which evolutionary explanation replaces the traditional "final cause" (108).
"The phrase 'evolutionary cause,'" Plotkin asserts, "can be sub-
stituted for 'ultimate cause,'" and evolutionary cause means, of
course, adaptation for the purpose of survival and propagation.
Explanation stops at the point at which the adaptive function of
the variation is identified, and thus insofar as things like consciousness or behavior can be understood as adaptive, their nature is explained, for if a phenomenon can be shown to be adaptive in this sense, it is thus validated. Things are what they are
because their natures are confirmed by their adaptability.

So, on the one hand, Darwin has provided occasion for an almost deconstructive undermining of basic patterns of Western
thought, what John Dewey gleefully described as "laying hands
on the sacred ark of absolute permanency" and "a mode of thinking that in the end was bound to transform the logic of knowl-

edge, and hence the treatment of morals, politics, and religion."[28] On the other, that mode of thinking left a gap that his
own theory has been called upon to fill. If "natural selection" replaces creation, then "natural selection" becomes the only available grounding concept for both scientific and political positions.
Just as "natural selection" became, in Darwin's first metaphorical
formulations, an agent who is "daily and hourly scrutinising"
all organic variations, so in current antimetaphysical discourse it
becomes an "algorithm" through whose formulae all organic life
passes. For both Darwin and Daniel Dennett, natural selection is
not literally an agent, but for both in effect it does the work of
God.[29] A demetaphorized (or remetaphorized) God emerges in
the most hard-nosed contemporary advocates of the mindlessness of nature, and this "power" becomes indispensable in re-
building nature from scratch. All true Darwinians can agree that
natural selection requires a "struggle for existence," but, as I
have already suggested, that phrase can mean many things, including cooperation, and the varied history of uses of Darwin
spins around how the word is interpreted. Even "natural selection," as Niles Eldredge argues, is no stable concept: "all Darwinians affirm that natural selection underlies adaptive change,"
but "how that motor actually works . . . is clearly something else
again."[30]

Whichever version of "natural selection" one takes, Darwin's
story of origins has become an inescapable alternative to "In the
beginning. . . ." Virtually any plausible theory of human behavior and organization must move through it. Biological explanation seems to provide an unarguable, material terminus—the
last elephant in that long tower of elephants.

But the question of foundations, the last elephant, is only partially epistemological. Finding a foundation is finding authority,
and we have already seen examples of the way in which Darwin's foundational arguments have directed lines of research
that reverberate with ideological implications. Foundational confidence often provokes premature plunges into social and political analysis, sometimes with awful results. Eugenics is only the

most obvious and terrible of such developments, virtually invented by Darwin's cousin, Francis Galton, but built on a reading of Darwin and a strong confidence that science provided the best means of working toward the social improvement of mankind. I have been trying here to face the reality that Darwin's writing can yield, with at least partial legitimacy, this kind of interpretation, and it is necessary to look a bit more closely and directly at how it does.

II

I am attending here and throughout this chapter to certain dominant interpretations of Darwin that, important and obviously relevant to his work as they are, point directly toward Weberian disenchantment, after all. I do not want to deny the validity of the interpretations, nor to suggest that Darwin didn't at times endorse some of the ideas that have given Darwinism a bad name among many cultural critics and historians, but I do want to suggest that these constitute only a small part, and often the intellectually weakest part, of his work. That Darwin at times seemed to sanction something like eugenics, that his own reading of the significance of his arguments became fodder for social Darwinists does nothing to belie the historical contingency of these arguments or the fact that they are not *necessary* consequences of his theories. By looking first at Darwin's own interpretation of his theory, and then, in later chapters, at some very different positions that have been defended as Darwinian, I want to develop and illustrate the argument that I have made in both the preface and the first chapter of this book: what Christopher Herbert described to me in conversation as "the ideological indeterminacy of ideas," and what Oscar Kenshur calls "ideological essentialism."[31] Not the absence of relation between ideology and idea, but the indeterminacy of the relationship. Darwin's theory is too complex to be affiliated simply with particular political models. In rejecting, for example, the usual connection of

Darwin's theory of "struggle" with endorsements of war, capitalism, and violence in human society, Paul Crook insists that we must not forget "the subtlety of Darwin's mature view of struggle."[32]

When in the *Descent* he took the plunge, arguing publically that "man must be included with other organic beings in any general conclusion respecting his manner of appearance on earth" (1:1), Darwin inevitably entangled himself in anthropology, sociology, and the human sciences in general, with all his ideological baggage exposed. He risked overtly putting into play as science his fundamental attitudes toward "cultural" rather than scientific subjects, toward race, class, women, free enterprise, to take the most obvious examples. And here, from my point of view, is Darwin at his worst, worst because much of the intellectual rigor and dependence on carefully investigated evidence seems, on these issues, to have been at least momentarily lost, and worst because the positions he takes are so blatantly based on unconsidered assumptions about the superiority of his own culture and class. In effect, the sequence from *The Descent of Man* that I want to look at here is his *interpretation* of the cultural implications of his own theory, and alas, he talks in ways that justify interpretations that led to eugenics and other more or less horrendous uses:

> With savages, the weak in body and mind are soon eliminated; and those that survive commonly exhibit a vigorous state of health. We civilised men, on the other hand, do our utmost to check the process of elimination; we build asylums for the imbecile, the maimed, and the sick; we institute poor-laws; and our medical men exert their utmost skill to save the life of every one to the last moment. There is reason to believe that vaccination has preserved thousands who from a weak constitution would formerly have succumbed to small-pox. Thus the weak members of civilised societies propagate their kind. No one who has attended to the breeding of domestic animals will doubt that this must be highly injurious to the race of man. It is surprising how soon a want of care, or care wrongly directed, leads to the degeneration of a domestic race; but excepting in the case of man himself, hardly any one is so ignorant as to allow his worst animals to breed. (*Descent*, 1:168)

The cultural prejudices at work in this passage are too blatant now to require analysis, but surely for Darwin—and this is certainly part of what makes passages like this so chilling—he was merely reporting the facts, being as objective as possible. He knew well the way breeders worked and tried his hand at pigeons himself, and it was merely reasonable to argue that breeders would never let their worst animals breed. As the subject of Darwin's work shifts from animals and plants to humans, a sentence like this suddenly carries enormous ideological weight. The absolute analogy between weak and vulnerable humans and "worst animals," not argued but taken for granted, changes what, in the *Origin*, would simply be a step in the larger argument toward natural selection into a moral reprimand and an implicit injunction.

The passage provides, perhaps, a too easy example of how ostensibly descriptive language is laden with cultural assumptions— clearly, any time the human becomes the object of study, everything reverberates with values, assumptions, ideology. What might be called objective, and would therefore be laden with authority, is the contrast between the savages out there and the civilized "we"; in metaphoric silence, the civilized are juxtaposed to the "savage," and society's weak are linked to the animals at the end of the passage, and both are regarded as Darwin's breeders of pigeons or dogs might regard them. It is just the attempt to be impersonal that makes it so chilling: "They propagate their kind." "The process of elimination" is, of course, the work of natural selection, and applied to the human, clearly it would be simply a killer.

It is important to recognize, however, that the factual statement is already an interpretation as well. It implies a simile: the "weak" in society, who need vaccinations, medical attention, or some forms of state support, are analogous to the worst animals; the breeders who wouldn't breed weak animals are analogous to the strong, who in fact run society. While it is clear to breeders of pigeons, seeking longer bills or puffier breasts, what the "worst" might mean, Darwin provides no cautions about the meaning of "worst" as he shifts the subject to humans. The unquestioned category "worst" requires a judge and a set of criteria, but this passage is written as though the identity of both were clear. Analogy, which does so much creative work in Darwin's writing, here in effect disguises moral decisions. Darwin leaves out of possible consideration the ways in which cultural factors might produce "weakness," or the effect of chance and contingency on poverty and illness.

The thinking is, if one looks back to the *Origin*, or even to the large part of the *Descent* in which Darwin discusses sexual selection, uncharacteristically slovenly. The care and attention that Darwin lavishes on bees, barnacles, or worms gives way here to relative hearsay from cultural commentators, many of whom might in fact have been influenced by Darwin's earlier thought, and the breeding simile that gives energy to this passage in fact becomes part of an extended interpretation.

But even here, where eugenics would seem to follow as the night the day, it is not the way Darwin chose to go. Darwin never became a eugenicist; for him instinctive human sympathy, "tender," "widely diffused," not only sustains these humanitarian activities, but through that sympathy avoids "deterioration in the noblest part of our nature" (*Descent*, 1:169). One version of rational consistency leads to a merciless closing down of aid to "inferior" members of society; another version sees sympathy, a product of animal "social instincts," as valuable to human society. If, he goes on to say, "the aid which we feel impelled to give to the helpless" is an "incidental result of the instinct of sympathy," it is one of the distinctive marks of the human. This is, then, not merely a matter of irrational and intellectually misguided compassion, it is a matter of interpretation, of determining which value has priority. Interpreting the facts he gives us, including the fact that breeders wouldn't let their weakest specimens breed, Darwin opts for compassion, if without much enthusiam. An entirely "rational" engagement with the facts might yield another conclusion. But even without the benefit of William James, Darwin interprets his materials in a way that transcends the "merely" rational, or perhaps to put it more carefully, in a way that incorporates feeling and desire into the fact.

The passage continues, however, and it is clear that whatever compassion might do for the "civilized" man, Darwin remains committed to the view that the weak propagate too much and threaten to dilute the world population and ultimately to triumph over the best. On the verge of evolutionary ethics, Darwin consoles himself throughout the passage with the fact that "malefactors are executed. . . . Melancholic and insane persons are confined, or commit suicide. Violent and quarrelsome men often come to a bloody end" (1:172). Similarly, "Profligate women bear few children, and profligate men rarely marry. Both suffer from disease" (1:173). This is the good news in civilization's efforts to head off the bad effects of its sympathy and generosity. Yet Darwin fears that W. W. Greg is right, that "the reckless, degraded, and often vicious members of society, tend to increase at a quicker rate than the provident and generally virtuous members," and that "In the eternal 'struggle for existence,' . . . the inferior and *less* favoured race" will in the end prevail. Trapped by his own compassion, Darwin has to content himself with the hope that "the weaker and inferior members of society" will not marry "so freely" as the sound (1:169).

Nothing in the theory of descent by modification through natural selection entails this kind of social analysis and program. When cultural assumptions are unquestioned, natural selection can be slotted in to do all kinds of work, and when social structures are taken as natural ones, in the sense that hierarchies within it are understood as analogous to or the same as hierarchies in uncivilized nature, natural selection stops being a metaphor. While Darwin, for example, recognizes that wealth does not inevitably equate with natural superiority, he complains primarily about civilization's aristocratic tradition, suggesting that primogeniture is a problem, since eldest sons, even those "weak in body or mind," marry and produce offspring. Yet again, the capitalist model is reinvoked; the evil consequences of primogeniture are at least partially checked because "men of rank" always wish "to increase their wealth and power" (1:170). At times, when Darwin writes of these issues in relation to current social conditions, the *Descent* begins to sound like a set of capitalist old wives' tales. We learn, for example, and what a relief, that Mr. Galton has shown that "the daughters of parents who have produced single children, are themselves . . . apt to be sterile" (1:170).

When Darwin played with his own flowers in his garden at Down, when he tested out breeding questions on his own pigeons, when he sought advice of botanists like Joseph Hooker, or bird experts like John Gould, he tended to be rigorous and cautious. In the sections on sexual selection in the *Descent*, one feels the tightness and rigor of the evidence and the argument, as one felt it in the *Origin*. Darwin carefully and exhaustively moves up the scale of life (here too, despite his occasional unhappiness with the words "higher" and "lower," he follows a hierarchy of life up to the human), from crustacea to spiders, to other insects and butterflies, to fish, amphibians, and reptiles, to birds, and then mammals, and finally "Man." Until he gets to the human, and particularly to the "civilized" human, Darwin's observational and ratiocinative powers lead him to places that are not contemporary commonplaces. As I will be discussing it in a later chapter, he can recognize what most of his contemporaries refused to credit, that among birds, for example, females have a power of choice that significantly affects the morphology of the species: "the exertion of some choice on the part of the female seems almost as general a law as the eagerness of the male" (1:273). The budding sciences of ethnology and anthropology did not, for the most part, serve Darwin well. So when he discusses human behavior that he hasn't studied exhaustively himself (it is different, for example, with the expression of emotions, for which Darwin's researches prepared him carefully), Darwin is not alert to the degree to which he is taking things for granted, the degree to which his informants are themselves taking things for granted, or are even uninformed. Talking with scientific authority about humans is a dangerous business, and while Darwin is on very strong ground when detecting in the life of animals traits that are distinctly human, he is on much weaker ground when he moves into the social sciences, whose development he so much influenced. The always hazy border-line between fact and value, between science and culture, be-comes, for the most part, too blurred to be functional.

Moreover, as we listen to Darwin's language when he treats the human subject, there seems to be a change, even when the formulations sound familiar. When Darwin tended to his plants, growing in profusion in his garden or a teacup, when he consid-ered the anatomy of barnacles, the intelligence of worms, the ar-chitecture of beehives, the structure of the eye, he manifests an evident enthusiasm and even awe, mixed with an observational acuteness that is absorbed rapidly into a larger argument. But when he writes about humans, usually with great caution, there are inevitable resonances, even as he continues to try to sound as objective as possible. "I know no fact in natural history more wonderful than that the female Argus pheasant should be able to appreciate the exquisite shading of the ball-and-socket orna-ments and the elegant patterns on the wing-feathers of the male" (2:400). The wonder is at the bird, and her subtle powers of aesthetic appreciation. But such subtlety is commonplace among humans and would be expected, with the result that the wonder implicit in the prose seeps out. When, a few pages later, Darwin notes that humans are "impelled by nearly the same motives as are the lower animals," he goes on to suggest that our breeding habits ought to be similar as well: "Both sexes ought to refrain from marriage if in any marked degree inferior in body or mind" (2:403). The difference is stunning. The leap from the "is" of the animal to the "ought" of the human does seem to vio-late Darwin's usual method of rational argument; his analogy has gone out of control. The descriptive language (transformed occasionally, as here, into prescriptive) is loaded with a kind of inverse affect. The objective language sounds as though Darwin is deliberately avoiding the affect that discussion of human behavior, most particularly sexual behavior, conveys. What results is something of a chill.

Near the end of the section of the *Descent* that argues the family relation between humans and the lower animals, there is a passage that interestingly suggests the range of affect in Darwin's writing. Here he is forced to confront what he knows will be the culture's unhappy response to the news: "Thus we have given to man a pedigree of prodigious length, but not, it may be said, of noble quality." But one of the marks of Darwin's writing and relation to nature everywhere is his enthusiasm for just those "lower" beings, a long way from what he rather embarassingly calls "Man, the wonder and glory of the Universe." The true wonder, as one feels it in Darwin, is the wonder of nature itself, and in arguing that we should not be "ashamed" of our parentage, he concludes with the voice of the enchanted naturalist: "The most humble organism is something much higher than the inorganic dust under our feet; and no one with an unbiassed mind can study any living creature, however humble, without being struck with enthusiasm at its marvellous structure and properties" (1:213). There is something conventionally complacent about the enthusiasm for "Man," and something authentically Romantic about the enthusiasm for our prehuman parentage. Locked as he is into a context in which his overall argument requires some apology to a culture whose view of the world and of itself has been by and large very different, Darwin lapses into lip service to Man and the Creator, but his energies and highest intellectual powers are reserved primarily for flowers, barnacles, spiders, and birds.

Moreover, on occasion, the real stakes in the human subject emerge, and by the end of the *Descent*, Darwin in fact rises to something like anger, pursuing the logic of his is/ought leap: "When the principles of breeding and of inheritance are better understood, we shall not hear ignorant members of our legislature rejecting with scorn a plan for ascertaining by an easy method whether or not consanguineous marriages are injurious to man" (2:403). Here, the intensity of Darwin's interest is obvious. He himself, having married consanguineously, constantly worried about the effect on his children (and with some reason), and the threat that every scientific "fact" about the human will have deep affective resonances is vividly exemplified. Darwin gets passionate here, not out of awe at the extraordinary powers of nature, but out of fright at the consequences, personal and social, of certain human actions that differ, on the whole, from those of the nonhuman organisms he otherwise studies. By the end of *The Descent of Man*, Darwin has moved into another mode, taking his "facts" and turning them into social policy. "There should be open competition for all men; and the most able should not be prevented by laws or customs from succeeding best and rearing the largest number of offspring" (2:403–4).

Partly from instinctive compassion, Darwin did not become a eugenicist, but not because he didn't think it was a good idea in principle. He thought of eugenics as "Utopian." Darwin is not happy at the prospect of the weaker members of society breeding freely and so urges voluntary abstention: the "weaker and inferior members of society" should not marry "so freely." Famously, Darwin suggests, "all ought to refrain from marriage who cannot avoid abject poverty for their children" (2:403). But he has no illusions, doesn't expect that sort of restrained behavior, and so argues that we must put up with the consequences of our compassion and tenderness.

The "application" of scientific ideas to social decisions entails the engagement of a whole person (perhaps Attridge's "idiogram"), and the extension of science to society is dangerous not only practically but individually. It requires a recognition that there are no innocent facts. Social management means making life and death choices. Even supposing that the whole theoretical backdrop were basically correct about the facts, the facts slip away into judgments (as in the word "inferior") without any indication that there has been slippage. They might have been interpreted differently, and even here, where Darwin sounds like a social Darwinist, he himself interprets them in two different ways.

The Darwin who emerges from *The Descent of Man* is a eugenicist who rejects eugenics. He is a celebrant of the natural world who is awed and enchanted by the complexity, intelligence, and variety of the lower animals. As he turns to interpret his own insights, to take what he has discovered of the nonhuman organic world and apply it to human beings and human society, all the assumptions of his culture rise to the surface. If it is true, as Desmond and Moore argue, that Darwin's theory of natural selection was not only closely connected with Malthusian economics but a means of reconciling this potentially revolutionary idea to the Whiggish culture to which Darwin belonged, the implications of that connection are overtly clear in Darwin's inferences about human social organization. This aspect of Darwin's thinking, his own interpretation of certain aspects of his own work, is not the aspect with which, for the most part, this book is concerned. The "model" Darwin I invoke in the later chapters of this book is the enchanted one, the naturalist who found in nature the emotional and spiritual resources to give value to life, the sublime multiplicity, complexity, and unique individuality that creates that "mood of fullness, plenitude, or liveliness" Jane Bennett identifies as characteristic of modern enchantment (4).

But like Bennett, I find the "charge of naïve optimism" a serious one: "it raises the question of the link between enchantment and mindlessness, between joy and forgetfulness." There can and should be no forgetting the social Darwin we have just been reading. There is, however and after all, the other Darwin, whose enthusiasm for female peacocks we have also just seen. But I agree with Bennett that "in small, controlled doses, a certain forgetfulness is ethically indispensable" (10). In the face of the Weberian narrative of disenchantment, in which the absence of a "divine creator," of a teleology, expels meaning from the world and leaves it barren, such controlled forgetfulness is particularly important. Bennett argues that "the good humor of enchantment spills over into critical consciousness and tempers it, thus rendering its judgments more generous and its claims less dogmatic" (10). More simply, the possibility of a naturalistic enchantment emerges as a crucial alternative to supernaturalist religion, which does so much harm when it imposes its norms on a secular polity.

Richard Dawkins claims, in a recent collection of his semipopular essays, that he is, as scientist, "a passionate Darwinian, believing that natural selection is the only known force capable of producing the illusion of purpose which strikes all who contemplate nature." But, he goes on, "I am a passionate anti-Darwinian when it comes to politics."[33] Dawkins is Darwinian in that he believes Darwin was right about natural selection (and in this respect he is more Darwinian than Darwin), and he is anti-Darwinian in that, like Huxley at the end of his prolegomena to *Evolution and Ethics*, he believes that we can rebel against the "unwelcome message of the Devil's Chaplain . . . the historic process that caused you to exist is wasteful, cruel and low." But that process has "blundered unwittingly on its own negation" (11). Dawkins thus separates a moral program from what he takes to be the scientific fact, although it must be based in part on knowledge and an understanding of the constraints that nature imposes on us. But even working "as a scientist" and rejecting the idea that what he learns as scientist imposes upon him a certain responsibility for political action, Dawkins is immediately sensitive to the way that Darwin has a "politics."

This brings me back again to the point that his theory does *not* have a politics; it grew within a form of politics and sustained itself both within that politics and outside of it. Politics, of many varieties, accompany it wherever it goes. But Dawkins, for example, is a long way from the Whig ideology that dominated Darwin's view of the world and indeed, in condemning the politics of natural selection, attempts to remain Darwinian while separating himself from nineteenth-century laissez-faire thinking. Readers and scientists create a politics as they read Darwin, but it is not as though they find their politics *in* Darwin.

Dawkins, in resisting the social-Darwinist reading of Darwin,

which, we have seen, on occasion Darwin himself offered, is satisfied that the sheer joy of reason, the power of its truth-bearing
energy, will be enough to do the moral work he wants to see
done. The very modern and implacably rationalist Dawkins
seems like a W. K. Clifford of our times. He sounds remarkably
like his Victorian predecessors, not only like Huxley, but like
those ironic and tough-minded scientists and intellectuals who,
in their powerful (and affectively impressive) rhetoric, unyield-
ingly insisted on the impossibility of knowing anything beyond
what could be scientifically affirmed. Dawkins takes no account
of William James's rejection of the thin gruel of these thinkers,
His move is a total, unmodified affirmation of scientific reason
and has about it the same air of maverick charm and eccentricity
that made Clifford popular. Dawkins is also, by the standards of
the dissemination of scientific ideas, popular. But against the
broad cultural resistance to affectless scientific explanation, he
remains ineffectual, and he does nothing to address the work
that Connolly insists is central to all human thinking: the work
of the "amygdala," without attention to which no ideas can have
their full impact.

The enchanted Darwin may well be another thing. Dawkins
sees himself as Darwinian in affirming the power of natural selection, and anti-Darwinian in asserting the necessity and possibility of moral resistance to its methods. To look at it that way,
however, is to assume that the only possible political interpretation of the theory of natural selection is social Darwinist, when
there are many other possible ones that history has thrown up
along the way. We need to stress also, as Dawkins does, that
nature's processes are not moral injunctions. Dawkins rightly
insists that natural selection has, in its sloppy, trial-and-error,
algorithmic movement, blundered "unwittingly on its own
negation." It has created creatures who live in culture, and culture has its own energies and powers that paradoxically can resist natural selection's pressures and procedures. Creatures like
us are in the odd position of being able to reflect on our own
genesis and to seek ways to control our futures, to imagine an

astonishing variety of possible significances of natural selection and evolution, and to build conflicting political and moral programs on those significances. We can, that is, regard with affect the ruthless, mindless processes that have produced us, and thus recognize with awe, admiration, and anxiety the miraculous richness of the world natural selection has helped to create.

CHAPTER 3

Using Darwin

I have focused thus far on the interpretations of Darwin that emphasize his indebtedness to his own class and moment. But historians of science have been aware of the remarkable variety of interpretations to which Darwin's ideas have been subjected, and if anything can suggest how impossible it is to consider any particular ideological positioning as *intrinsic* to a theory such as Darwin's, that variety should do it. Janet Oppenheim argues that "Darwin's theory of evolution was infinitely pliable. It could be twisted to justify militarism and pacifism alike, imperialism and cooperation, unbridled laissez-faire capitalism and socialism. Perhaps the chief reason for the ubiquity of evolutionary modes of thought . . . lay expressly in the capacity to appeal to all ideologies."[1] "Darwinian science," says Paul Crook, "was multivalent . . . it inspired liberalism as well as reaction" (199). Or as Steve Jones has recently put it, "Evolution has been an alibi for socialism, for capitalism, and for racism; and no doubt would have been seized on by the one hundred thousand systems of belief that [E. O.] Wilson estimates have existed since consciousness began."[2] Darwin was almost all things to almost all men—and women—or, as George Bernard Shaw claimed rather contemptuously, "He had the luck to please everybody who had an axe to grind."[3]

But some uses are more Darwinian than others. It is difficult to disentangle, from among those who are committed to the idea of evolution by natural selection, those who are pro-Darwin and those who are anti-Darwin. The popular view of Darwin regards as Darwinian any of the ideas that either rely on Darwin's theory in any way, or that *seem* to rely on them, as Herbert Spencer's do, no matter how different Spencer's thought was from Darwin's. Certainly, whatever the version of evolution taken, its cultural power depends on the fact that Darwin became, for complicated historical reasons, the figure who is somehow responsible for the idea of evolution per se. Darwin really is everywhere, and among those who are committed to evolution as a secular explanation of life and its origins, there is often a contest to be more Darwinian than thou. Steve Jones's word "alibi" suggests one of the major points underlying my arguments through this book: that Darwin's theory is more guilty as an excuse for some ideological bias or other than as an adequate, logically coherent cause of that bias. This is not to suggest that Darwin himself did not interpret his ideas in the direction of certain social objectives, as we saw in the last chapter, but only to insist that whatever the roots of the theory (in Malthus particularly), it can be taken almost anywhere to do ideological and political work.

Since nonscientific uses have been the norm not the exception, social theory has seemed to need Darwin. For many scientific theorists today, social theory needs to become more Darwinian, more "scientific," a tendency that carries its own burdens of ideological interpretation, to which I will be referring in the next chapter. What needs particular critical attention is the way Darwin has become the foundation for scientific projects that continue to generate ideological applications. Of course there are—or there should be—constraints on the possibility of interpretation, and we have already seen that there have been dominant directions in the uses of Darwinism. But logical constraints have little to do with the life of the idea in culture; moreover, as Diane Paul repeatedly points out, "Darwin's waverings" on social issues have contributed significantly to the multiplicity of ways in which he has been used.

Natural selection is a radically materialist reading of the world's processes, and it would seem to require the greatest ingenuity to use it in support of religious views—yet, this too has certainly been done.[4] Natural selection also finally requires death for life, and Darwin's description of it strongly emphasizes—there is Malthus behind him—a constant struggle for limited resources. This of course provides the prima facie connection with political theories that privilege competition—but this connection, too, hasn't always been made.

The "mechanism" Desmond and Moore describe as "compatible with the competitive, free-trading ideals of the Ultra-Whigs" was also a mechanism that deeply upset some of these same free traders, that threatened fundamentally the sorts of ideological stability that Malthusians sought, that appealed in some of its aspects to anarchists and socialists and pragmatists, and of course to future eugenicists and libertarians and industrialists and evolutionary psychologists (who seem, among themselves, to have different political affiliations).

There are those who resist the almost unending proliferation of interpretations to argue that after all nature did it, that the revolutionary in Darwin emerged from his direct engagement with the scientific problem rather than from his preoccupations with social issues, or that his theory is ultimately independent of whatever ideological tendencies helped him to it and simply registers what is true.[5] Another view, provided by Desmond and Moore themselves, in arguing against Ted Benton's insistence precisely that nature itself had to be the determinant, is that the "revolutionary" element in Darwin's argument reflected the complexity, multiplicity, and contradictoriness of Darwin's "circle." Benton's fear is that our current preoccupation with emphasizing the ways in which scientific ideas are implicated in society and culture can lead to undervaluing scientific knowledge itself. The debate over this issue—the degree to which it is nature, not culture, that determines how scientists work—has been heated and extended, but Peter Bowler is surely right that "the theoretical structure of any modern science is far too complex to permit finding one-to-one correspondences between its concepts and those wider issues everyone must confront."[6]

Nevertheless, a long tradition of applications of Darwin claims one-to-one correspondences, and I want now to look at a few such applications that might be seen as representative, assuming the priority of science in all areas of knowledge, building on the extension of science to the organic and the human, and turning it immediately into ideology. The variations here should suggest how hopeless it is to make that one-to-one connection, to stabilize Darwin's meanings in cultural matters and hook them intrinsically into particular political positions. This is not so much an attempt to exonerate Darwin from the applications of his theory that I don't happen to like as to open space for my own alternative sense of how one might find the "enchanted" Darwin and how one might best "use" him.

I will take my leading example from the interesting hostility of Karl Pearson to Benjamin Kidd. Both were late-nineteenth-century writers who may be understood to represent main currents of response to Darwin and who anticipate the kind of developments I will be discussing when I turn, in conclusion, to Steven Pinker and E. O. Wilson. Both Pearson and Kidd might be considered "social Darwinists," and that label is useful here because in its application to these two it becomes clear that even "social Darwinism" is not one thing. Kidd's *Social Evolution* (1894) was a popular application of Darwinian theory to social analysis. Pearson, whose *Grammar of Science* (1892) made a brilliant case for rethinking scientific knowledge and for seeing it as both a moral and epistemological enterprise, was a strong socialist, an almost but not quite complete feminist. Though influenced by German thought, particularly Kant and Ernst Mach, he often, in his celebration of science and free thought, sounded a great deal like the scientific naturalists. His long-term effort to extend the dominion of science into the study of human behavior took the shape of statistical analysis and kept *The Grammar of Science* in print and in an "Everyman" edition for at least fifty years. For Pearson, making the social sciences truly scientific was a necessary condition for the project of eugenics, which he, polymathic disciple of Francis Galton, took as the best hope of a humanity that was subject to the processes of evolution and natural selection.

The fundamental point of agreement between Kidd and Pearson is the centrality of scientific explanation to understanding (and controlling) human behavior.[7] Kidd's project—like

Pearson's, though it emerges from what seemed to both of them utterly hostile political positions—interestingly foreshadows sociobiology, for, he claims, "all departments of knowledge which deal with social phenomena have their true foundation in the biological sciences."[8] To understand the advance man has made (and his book is full of the wonder of that advance) one must understand that natural laws have operated to produce it. Social systems, he claims, are "organic growths . . . apparently possessing definite laws" (31). Kidd concludes his opening chapter with this manifesto:

> The time has come, it would appear, for a better understanding and for a more radical method; for the social sciences to strengthen themselves by sending their roots deep into the soil underneath from which they spring; and for the biologist to advance over the frontier and carry the methods of his science boldly into human society where he has but to deal with the phenomena of life where he encounters life at last under its highest and most complex aspect. (28)

Pearson makes even larger claims for science. So, at the start of *The Grammar of Science*, he insists that

> modern science does much more than demand that it shall be left in undisturbed possession of what the theologian and metaphysician please to term its "legitimate field." It claims that the whole range of phenomena, mental as well as physical—the entire universe—is its field.[9]

Later he develops the claim further, arguing that we should "recognise that science can on occasion adduce facts that have far more *direct* bearing on social problems than any theory of the state propounded by the philosophers" (25). Like most of their contemporaries, then, and like many of our own, Kidd and Pearson both believed not only that there is a complete continuity between the physical and the human sciences but that it is the responsibility of serious thinkers about society to recognize the truths of biological science underlying mental and social phenomena. Although their premises were linked, their differences obviously emerge from the different ways in which they interpret the workings of natural selection. These attitudes, it should be added, would seem also consistent with Weber's analysis

of the ways in which science, extending its reach to explain everything, further disenchants the world. But at the same time—and
this is key for my overall argument about secular enchantment—
both of them see the science they are invoking as a system growing out of and allowing the greatest use of values and ethics.

The disagreements between Kidd and Pearson cannot be easily assimilated to our now conventional distinctions between
"right" and "left." Pearson attacked Kidd in the name of socialism
and true Darwinian science; Kidd fought for what he thought of
as liberal democracy, imagining through the war of each against
each some ultimate altruism in the process of natural selection.
Where Kidd argued that natural selection would work more
gently than many interpreters of Darwin suggested, and thus
could produce a satisfactory social order, Pearson, as Theodore
Porter points out, believed that "scientific planning was . . . the
next phase in human evolution" (202). Both Pearson and Kidd
were anticapitalist, hostile to the ruthless and destructive competition that dominated late Victorian society, and yet both were
imperialist in their understanding of the relation of the white
European to nonwhite races.

These writers are representative in part because they make
clear what has been persistent in the uses of Darwin since the development of the theory: that altruism is a problem.[10] Darwin recognized that directly and in chapter 3, in *The Descent of Man*, he
tries to show at some length that "the most noble part of our nature" derives not from "selfishness" but from the social instinct.
Although the "selfish gene" may have been hiding in Darwin,
waiting to get out, Darwin was not himself comfortable with a
notion of the natural universality of selfishness. The problem first
enters Darwin's work in the famous chapter on the "struggle for
existence," over which contending ideologies have themselves
struggled for a century. There the phrase "struggle for existence"
is used "in a large and metaphorical sense" (*Origin*, 62)[11] and includes the phenomena of interdependence and mutual aid. In the
Descent, Darwin would argue that "those communities, which included the greatest number of the most sympathetic members,

would flourish best and rear the greatest number of offspring" (1:82). The raw Spencerian notion of survival of the fittest, a turn of phrase not Darwin's own but which he was persuaded to borrow, does not take altruism into account, but any political theory that tries to get beyond rampant individualism must allow for altruistic behavior. Pearson and Kidd contend over that battlefield, both assuming that political theory requires scientific support and both looking to Darwin to provide it.

Kidd's science would seem at first all too familiar to those who make the McGinnis-style connection between "Darwinian" science and an emphasis on universal competition. Adopting a tone characteristic of our own contemporary sociobiology and reductionism, Kidd wants us to face "the stern facts of human life" (58). Natural selection does its work regardless of the most generous and merciful interventions, so while "the Anglo-Saxon looks forward . . . to the days when wars will cease," he is nevertheless "involuntarily exterminating" all the nonwhite races with whom he comes in contact. This notion of extermination, looked on with approval, if with some regret that the laws of nature require it, was characteristic of late Victorian imperialist thought and was strongly affirmed (and advocated) by Pearson. The processes of nature are inexorable, and if there is to be progress it will happen in accordance with those processes: "natural laws operate in producing the advance man has made in society" (30–31), and thus good intentions toward weaker peoples are no less destructive of them than bad ones. This is a familiar ploy of late-nineteenth-century objectors to charity—an objection that came at times both from the "right" and the "left." But Kidd is *not* arguing for a policy of extermination—much the reverse. He is arguing only that the processes of natural selection will end by exterminating even as we follow our higher moral impulses to "regulate and humanize" the stern law. That is to say, natural selection does not determine morality; morality is often resistant to it. As we have seen already with Huxley, and with Dawkins in recent years, even the most convinced Darwinians recognize the need not to take natural selection as the standard for moral action, although Kidd acquiesces (sadly, he keeps reminding us) in the brutal work it effects in pruning out the less fit.

Surprisingly, however, Kidd's project is no familiar late-Victorian celebration of racial and class superiority and laissez-faire. He needs altruism in this competitive world and finds it not at the level of the individual (although the individual may be altruistic) but at the level of the species. He asks "whether religious systems have a function to perform in the evolution of society" (20).[12] Is religion a means by which the natural force of competitive violence is softened? Darwin himself had said that it was. If religion has played and continues to play an important role in the evolution of society, religion will, as Kidd says, "follow its course independent of our opinions" (20). Thus, ironically, the anticlerical propagandists for science are being unscientific, ignoring the "immense utilitarian function" of religion in evolution. But it is worth pausing to remind ourselves that such questions emerge from the assumption that all life, even human cultural life, is subject to scientific and naturalistic analysis. "The races who maintain their places in the van," Kidd claims, "do so on the sternest conditions. We may regulate and humanise those conditions, but we have no power to alter them" (58). The conditions that lead to the unintentional extermination of nonwhite peoples operate ruthlessly within advanced societies. "The extinction of less efficient forms . . . is the condition of progress" (38).

The thrust of this argument, though it is complexly made, is that *rational* attempts at equality in human society, attempts at justice and equity, socialism in particular, are doomed to failure. It is important to note the force of the word "rational." Kidd's point is that while the system of nature is ordered, it does not answer to the rational needs of individuals. To put it crudely, to be rational under the reign of natural selection is to be selfish; to be irrational is to curb one's selfishness, but only by thus being irrational can the species, as opposed to the individual, thrive. Rationality operates, but at a level higher than that of the individual. Socialism, which seeks rational satisfaction for all individuals, is thus a hopeless undertaking. For, says Kidd in italics, *"if all the individuals of every generation in any species were allowed to equally propagate their kind, the average of each generation would continually tend to fall below the average of the generation which preceded it, and a process of slow but steady degeneration would ensue"* (37). This echoes Darwin, who, in the *Descent*, it will be remembered, argued that "excepting in the case of man himself, hardly any one is so ignorant as to allow his worst animals to breed" (168). The pressure Kidd exerts on Darwin's theory, keeping his own "scientific," clearly emerges from a moral revulsion from the model of conflict and ruthless slaughter that he would seem bound to endorse. He argues that the interests of the individual will always be at odds with those of the "social organism," so that there "can never be found any sanction in individual reason for conduct in societies where the conditions of progress prevail" (80). What is required, then, for the best interests of the social organism and of social progress, is that man must "check and control the tendency of his own reason to suspend and reverse the conditions which are producing this progress" (82). And the work of such checking is the work of religion, which depends upon the irrational to restrain the individual's natural desire.

Interestingly, Kidd's turn to religion and his argument for the possibility of altruism depend on a view of natural selection that, until recently, had been discounted by most evolutionary theorists: group selection. Paul Farber points out how "Darwin and his contemporaries relied on the concept of group selection. A group that harbored altruistic individuals would have an advantage over those groups that did not." Through most of the last half of the twentieth-century, however, the concept of group selection has been out of favor."[13] The history of this development is compex, but several evolutionary biologists have recently argued forcefully for the possibility of group selection and the idea is reemerging as a scientifically serious one.

Most interestingly, David Sloan Wilson has argued, like Kidd, for the evolutionary importance of religion. Wilson does not accept the Victorian understanding of group selection, which

posits individuals willing to make large-scale sacrifices for the
group. Within any group of that kind, the sacrificing individuals
would be eliminated, and thus the group would cease functioning successfully against other groups. But Wilson argues that
higher-level selection can work because it need not entail "self-
sacrificial altruism," but relaxes "the trade-off between group
benefit and individual cost" (20). Religious groups, Wilson argues, "are based on much more than voluntary altruism." Wilson describes how "in the 1960's, adaptation at the level of groups
was rejected so strongly that the ensuing period could be called
the age of individualism" (6).[14] But the phenomenon of religion
entails another look at the question, and Wilson, sticking rigorously to the materials of evolutionary biology, but sensitive to
cultural developments as well, holds that "we should think of
religious groups as rapidly evolving entities adapting to their
current environments" (35). Kidd's group selection is a long way
from Wilson's, but both find that group selection helps explain
the evolutionary significance of religion.

Wilson, who takes religion seriously on his own terms, seems,
in moving toward group selection, to be moving away from biological determinism and some of the implicit social constraints
that have tended to go with it. Ironically, Kidd's group selection,
which also helps explain religion, tends to be socially much
more conservative than Pearson's Darwinism. But, like Pearson,
Kidd is no unrestrained free trader; he is in fact appalled by the
immoral, irresponsible focus on profit that dominates contemporary capitalism and that is sanctioned by the political theory
that he takes as the brilliant epitome of the progressive movements of modern history, that of the Manchester school. He argues that Darwin's own thinking does not sanction the wild
individualism of laissez-faire economics. "*Laissez-faire* competition," he says, "is, in the last resort, nothing more or less than . . .
a surviving principle of barbarism, necessarily tending, under
all its phases, towards the conditions of absolutism."[15]

Kidd rereads Darwin so as to point to a different kind of
restraint—a nonrational, nonsocialist restraint—built into the

evolutionary process. His point of departure, he claims, derives directly from Darwin: he sees evolution working on what Wilson calls higher-level groups; not all actions, he claims, are determined by the needs of the individual as opposed to those of the group. In his *Principles of Western Civilization*, he quotes unfavorably those passages from the *Origin* that point to what he calls "presentism," that is, the view that evolutionary action is always guided by the individual's relation to the present. "Any being," he says, quoting Darwin, "if it vary however slightly in any manner *profitable to itself* under the complex and sometimes varying conditions of life, will have a better chance of surviving and thus be naturally selected" (41). But such selection depends on the fact that natural selection "acts exclusively by the preservation and accumulation of variations *which are beneficial under the organic and inorganic condition to which each creature is exposed at all periods of life*" (42). Yet this merciless process is not what it seems, and there is space for altruism after all if one moves from the presentism of Darwin's arguments here. Kidd does not reject Darwin—he claims, rather, simply to be updating him—as he argues that evolution works by serving the interests of the greatest number of organisms. That greatest number can never be in the present; the majority is, rather, "the long roll of the yet unborn generations."

> other things being equal . . . the winning qualities in the evolutionary process must of necessity be those qualities by which the interests of the existing individuals have been most effectively subordinated to those of the generations yet to be born.
>
> It cannot, in short, have been simply the qualities useful to the individuals in a mere struggle for present existence which have directed the process of Natural Selection as a whole. (44)

Here is Kidd's version of group selection, although his idea that natural selection can work on a future not actively at work on the organisms of the moment seems very un-Darwinian and certainly antithetical to the Darwin accepted by modern scientists, whether or not they believe in group selection. For Kidd too, committed as he is to evolutionary explanation, altruism is

possible only if natural selection works on groups as well as individuals. Darwin, of course, though he held that morality derives from the social instinct that is part of herd behavior, always
insisted on individual variations, and always argued that natural selection never worked to the detriment of the organism that
it tends. Here, as virtually everywhere else, there are grounds for
disagreement about what the cultural—and, indeed, scientific—
implications of Darwin's ideas are.

Altruism, Kidd maintains, enters the system because the governing force in evolution is "the need of the species" rather than
"the molecular peculiarity unchangeably inherent in life" (49).
Altruism, action for the good of others that does not also entail
ultimate gratification for the actor, and within a system that is
moving inevitably in the direction of satisfying that future majority, allows Kidd to find a "meaning" in natural selection,
which he found otherwise to offer no meaning whatever. In
effect, Kidd imports teleology back into Darwinism (and thus,
one might add, makes it more obviously compatible with enchantment). There is, he says, a "principle of inherent necessity
in the evolutionary process compelling ever towards the sacrifice on a vast scale of the present and the individual in the interests of the future and the universal" (57). The most successful
"races" have been those that have most effectively subordi-
nated present desires to the "greater interest of their kind in the
future" (68). The great achievement of the British, then, has been
that in their military power, "representing the highest possibilities of militarism in the world," they have been "able to hold
the present for the future against all comers" (467).

For Kidd, then, a legitimate political reading of Darwinian
(and post-Darwinian) theory entails the affirmation of an imperialist liberal democracy constrained by religion to protect the
poor, to enfranchise unskilled labor, to recognize its right to a
minimum wage, to uphold its standards of life, and, generally,
"to enforce by law a class of claims representing in the last
analysis nothing more than the first bare conditions of free competition in its relations to capital on the one hand and to its own

kind on the other" (471). All of this activity is in the interest of allowing natural selection to play itself out most fully—given its teleology in the species, not the individual—in its entirely "rational way." Under the sway of processes indifferent to individual human needs, the species, the human majority, will progress. While it is difficult to reconcile Kidd's futurism with the natural selection that Darwin described, it is not unreasonable to argue that human culture has proved itself adaptive by transcending Darwin's "presentism" and working with an eye to the future.

The most remarkable twist in Kidd's "scientific" thinking, as he seeks to make natural selection both scientific and merciful, is his turn to religion. This depends on something like a reversal of the dominant interpretations of natural selection in human society. That is, he claims that "when man became a social creature his progress ceased to be *primarily* in the direction of the development of his intellect." In society, natural selection no longer works in cahoots with the Manchester school or with the Darwinian argument that all change emerges from the individual's engagement with the conditions of present existence. "His interests as an individual were no longer paramount," says Kidd, but became "subordinate to the distinct and widely-different interests of the longer-lived social organism." So in developed society, "large evolutionary forces" operate "through the instrumentality of religious systems," which "are securing the progressive subordination of the present interests of the self-assertive individual to the future interests of society" (285–86). Remarkably, Kidd can then go on to claim that natural selection is "steadily evolving in the race . . . religious character. . . . The race would, in fact, appear to be growing more and more religious." For 1894 this seems an astonishing claim, but it follows logically from Kidd's attempt to use a theory modeled on laissez-faire economics to do the work of social justice. The invisible hand turns out to function through natural selection.

Kidd's curious variations on Darwinian science did not allow him, as they did Pearson and most of his contemporaries, to accept either eugenics or the notion of the natural inferiority of the tropical races (as he put it). As Paul Crook points out, Kidd saw Pearson's eugenics as "authoritarian," and in 1904 was appalled that "while Galton proposed to reconstruct the human race by scientific breeding, he could find no place in his plan for moral standards," and in the end he came to believe that "Darwinism was both the flower of western science and 'the organised form of the doctrine of the supremacy of material force'" (Crook, 92). At the end of *Social Evolution*, Kidd anticipates the recent arguments of Jared Diamond, refusing to accept Galton's and the culture's assumption of the natural inferiority of "uncivilised races." Galton, he says, confuses "the mental capacity with which nature has endowed us" with "the mental equipment which we receive from the civilisation to which we belong" (271).

Although he is unlike some modern theorists of evolutionary psychology, Kidd insists on the distinction between cultural forces deriving from "civilization," and the hard-wired mental equipment with which natural selection has endowed us. But Kidd is a clear example of what I take to be the norm of Darwinian scientism in political and social application. That is, the determining factors of how Darwin's theory was to be used were, finally, the particular political commitments of the interpreters, commitments only fragilely dependent on the science they invoked. Finding a space for the altruism that virtually all agree is a distinctive character of advanced civilization is the crucial task, and finding a sanction for altruism in biology—almost a determinist necessity—obviously seemed the right way to move. Kidd's reformist and religious leanings clearly led him to make evolutionary science do the work of liberal democracy and of the culture's long-standing Hebraic tradition of—to reverse Matthew Arnold's term—not doing as one likes but subserving one's individual needs to the larger needs of the community.

Pearson attacks Kidd in the name of socialism and a more rationally understood altruism. He fights the battle too over Darwin's body, but also over the degree to which Kidd's arguments can be taken to be "scientific" at all. "It is open to question," Pearson condescendingly notes in passing, "whether Mr. Kidd has ever studied his Darwin" (121). If for Kidd Darwin justifies the institution of religion, for Pearson he justifies socialism. But Pearson is determined to be rigorous about science, to be empirical and, particularly, statistical, and he requires of all participants in the debate that they provide strong scientific evidence for their theories. So he complains immediately about the mistaken notion that "all that terms itself evolution must be scientific"[16]; much that passes for evolution is mere speculation. So the political battle will be fought over the question of what is scientific.

Pearson rather contemptuously cites reviews of Kidd's book that describe it as "an application of the most recent doctrines of science to modern society and life . . . an application of the laws of evolution announced in the *Origin of Species*." Bah, humbug. Confidently he dismisses Kidd: if Kidd's theory is correct, "the modern socialist movement is completely futile . . . opposed to fundamental biological truths" (107). But Pearson rejects all those readings of Darwinism that place it as a supporter of conservative domestic politics because they all fail to take account of the fact that the "social instinct" and "the altruistic spirit" are themselves the product of natural selection.

The Darwin Pearson reads does not mean that intraspecies conflict is the norm, nor that it is a condition of progress, as Kidd has argued. For Pearson, "the struggle for existence" must be analyzed carefully, and he distinguishes among "intra-group struggle, physical selection, and extra-group struggle." Kidd makes the mistake, says Pearson, of assimilating all of these into one conception of a Hobbesian war of each against all. Darwin's "argument as to the struggle for existence in plant and animal life is drawn," Pearson claims, "from the conception that we are dealing with a *practically stationary population*" (123). Applying those arguments, as Kidd does, to "the problem of the social evolution of civilized man" is simply a mistake. Without "any statistics and without any demonstration," Pearson objects, Kidd tries to apply to human society Darwin's thesis that only a "small number" of "the many individuals of any species . . . can survive" (*Origin*, 61); in so doing Kidd merely demonstrates his misunderstanding.

While Kidd makes his argument work by assuming group selection, Pearson's theory depends on an attempt to discriminate
more carefully between extra- and intraspecies competition: "The
particular factor of natural selection—intra-group struggle plays
little, if any, part among civilised man" (132). Since everything
for Pearson depends on whether the arguments are scientific, he
tries to support this claim with statistical analysis of such things
as mortality tables. He quotes rather selectively those passages
in *The Descent of Man* that emphasize the difference between the
development of man in civilization and animal development
elsewhere. Given the tentativeness of many of Darwin's arguments, there are plenty of passages to justify Kidd's reading too
(for example, Darwin says about the development of man up
from savagery that "it may well be doubted whether the most
favourable [variations] would have sufficed, had not the rate of
increase been rapid, and the consequent struggle for existence
severe to an extreme degree" [1:180]).

Nor does Pearson need religion, so necessary to Kidd, to ease
the savage competition of natural selection. Here is the culmination of his argument, as he justifies the socialism with Darwinian
science:

> While the socialist denies that intra-group struggle in civilised communities
> is ever to the death, he is quite ready to admit that intra-group competition
> may be of great social value, as putting the right man into the right place, as
> a means of obtaining a maximum of efficient social work. On the other hand,
> he holds that this competition can be carried on at too great a price; it may
> render the group unstable by the overwhelming advantages it gives to indi-
> viduals; it becomes disastrous the moment it approaches a struggle, not for
> comparative degrees of comfort within a limited range, but for absolute ex-
> istence. The socialist feels that in proposing to regulate this competition, he
> is not flying in the face of biological laws and cosmic processes, but taking
> part in the further stages of that evolution by which civilised man has been
> hitherto developed; this is just as much "biological" and "cosmic" as the
> evolutionary history of ants or bees. (130)

Reading Darwin into socialism, Pearson also reads out Kidd's religion. In fact, the socialist Pearson has rather a more hierarchical

notion of human society than even Kidd, for his explanation of the function of religion both accepts Kidd's analysis and rejects his conclusions. On the one hand, Pearson agrees that religion has served a useful function of increasing "social feeling at the expense of the individualist"; on the other, he sees religion not as currently growing but as being of use now only to the "unreasoning," for whom "the fear of future punishments and the hope of future rewards could have an effect" (115). This, he suggests, is hardly a satisfactory analysis for theologians, and virtually any other of Pearson's essays would make clear that the real project of the future is to supplant this unreason with scientific knowledge, knowledge that would confirm the validity of socialism. He argues: "That group in which unchecked internal competition has produced a vast proletariat with no limit of endurance, or with—to use a cant phrase—'no stake in the State,' will be the first to collapse" (131). Science thus supports both socialism and imperialism.

But Pearson's altruistic (if tough-minded) socialism depends on another level of struggle: extragroup struggle. Altruism becomes an instrument of extragroup competition. The more altruistic any given society is, the more powerful it will be in dominating "inferior" races and leaving a superior progeny.[17] So like Kidd, Pearson talks of how, "one after another inferior races are subjected to the white man," and the "stability and power of any group depends on the preservation and increase of its internal stability." One of the more chilling sentences in any of this literature comes when Pearson, shifting from "intragroup" to "extragroup" struggle, argues that "No socialist, so far as I am aware, would object to cultivate Uganda *at the expense of its present occupiers* if Lancashire were starving" (11).

Altruism, then, strengthens a society in its large-scale competition with other societies. Altruism is a competitive advantage. The more altruistic any given society is, the more powerful it will be in dominating the inferior races, which are partly marked by the absence of intragroup altruism. Competition and conflict reemerge for Pearson as for Kidd, though both try to resist the philosophy of the Manchester school while finding a way to read Darwin so that he sanctions their politics.

It won't do, in looking at these two figures, to imagine ways in which Darwinian thought aligns itself with conservative or progressivist thinking. For one thing, the very notion of "conservative" or "progressivist" is historically inflected, and while Pearson argues for a kind of socialism, it is easy enough to read him, too, as deeply conservative, even reactionary, in our contemporary terms. Kidd's commitment to laissez-faire economics acting itself out through natural selection would seem to fit neatly into our general sense of social Darwinism, except that his ideas are significantly shaped by hostility to rampant capitalism and the notion of "survival of the fittest." That is, in many ways, Kidd's capitalism is more "progressive," certainly more sensitive to individual needs and the responsibilities of democracy, than Pearson's socialism. These two figures are only two of an almost endless range of possible political twists to evolution by natural selection, and I focus on them merely as exemplary of the inadequacy of imagining some intrinsic political application of Darwinian theory.

Both turn what they see as the objective authority of science into a social program. In this respect, however distinctive and antithetical their approaches and politics, they are rather like dozens of others who read Darwin into social theory. And while there is no need to elaborate more of the curious ways in which Darwin becomes the ground for political positions he was not likely to have endorsed, it is worth pointing to another extreme that significantly weakens the role "struggle" plays in the work of natural selection. Of all the curious uses of Darwin, "anarchism" was perhaps the most curious. And yet the grounds for a Darwinian anarchism are, it seems to me, about as strong as for a Darwinian socialism or a Darwinian religion. Once again, the critical question is about the significance of the word "struggle," particularly as Darwin uses it in the chapter of the *Origin* called "Struggle for Existence." That chapter has all the earmarks of

Darwin's acquaintance with Malthus, and yet there are passages in it that turn the idea of "struggle" on its head and incorporate into it elements that might be called love. Plants on the edge of the desert are not so much struggling for life as "dependent on the moisture"; "missletoe is dependent on the apple and a few other trees, but can only in a far-fetched sense be said to struggle with these trees" (63). Struggle, in Darwin's use of the term, incorporates a wide range of relationships, only some small portion of which might be said to entail violent contest. Peter Kropotkin, who was a geographer and, as he thought of himself, "a scientific anarchist," read the Darwin that Huxley and the social Darwinists did not choose to see. Kropotkin's well-known book *Mutual Aid* reverts to that crucial passage in which Darwin widens the idea of struggle beyond that of conflict and competition, taking "struggle," as he says, in its "large and metaphorical sense." Kropotkin, in fact, uses many of the same arguments Pearson uses to focus on the way Darwin emphasizes the importance of sympathy within communities and quotes from the *Descent* the passage I've already cited about the way a society with the highest number of sympathetic members would flourish best.

As against the social Darwinists, Kropotkin insists "that no progressive evolution of the species can be based upon . . . periods of keen competition."[18] He complains that Huxley, Spencer, and in fact most commentators on Darwin "reduced the notion of struggle for existence to its narrowest limits," and in so doing "made modern literature resound with the war cry of *woe to the vanquished*, as if it were the last word of modern biology" (4). But Kropotkin too depends, even unself-consciously, on the assumption that science is the ground of social theory. If Darwin can be shown to have meant that "struggle" implied sympathy, then we have the logical right to extend the idea of sympathy in our own social theory. So he claims that "Huxley's view of nature had as little claim to be taken as a scientific deduction as the opposite view of Rousseau" (5). Huxley is not so much morally as intellectually wrong. He has got the science wrong.

If, as saintly anarchist, Kropotkin has not been deeply influential in modern political or scientific understandings of Darwin, his determination to find those elements in Darwin that treat "struggle" in terms of sympathy parallels the culturewide determination to ground politics in science. Kropotkin insists that his own worldwide researches did not confirm the social interpretations of Darwin that produced the vision of nature red in tooth and claw. That is, he found in nature the kind of evidence that he thought would justify another kind of social analysis. Nature for Kropotkin cries out, not "Struggle! Compete! Vanquish!" but "Don't compete!—competition is always injurious to the species, and you have plenty of resources to avoid it." So Kropotkin, Kidd, Pearson, from whatever different perspectives, in effect deny Gould's arguments that the "ought" of social policy cannot be derived from the "is" of Darwinian science. What they all agree on without having to make the argument is that "man" has become a legitimate subject for scientific study.

Of course, the range of possible Darwinisms extends well beyond these late Victorian interpretations. I am interested in these particular ones primarily because they make so abundantly clear how extravagantly opposed "Darwinian" positions can be. They all, of course, take solace from being "scientific" and they are all, of course, partial in their reading of Darwin. Their engagement with Darwin is impassioned; they enlist him in strong moral enterprises; they clearly do not read him as emptying the world of meaning. Instead, they take Darwin as a source—the world does not lose its wonder with Darwinian explanation; rather, it fills with meaning. This tendency in responses to Darwin has been pervasive and continuing, and as this book unfolds I want to make yet another use of Darwin, but one that self-consciously loads the world with meaning. It will be another partial Darwin, but one that can re-enchant the world.

CHAPTER 4

A Modern Use
Sociobiology

Having looked briefly at some late-Victorian uses of Darwin, and having attempted to suggest the degree to which Darwin himself bears the responsibility for these enterprises, I want now to consider some very recent, clearly and explicitly "Darwinian" considerations of the same problem: sociobiology and its immediate and apparently more effective descendant, evolutionary psychology. Taking scientific explanation deep into the human psyche, these enterprises would seem the last word in disenchantment, and the furious debates over their validity suggest that there is much more at stake here than whether their arguments are truly Darwinian.

Whether the disagreements are political or scientific is often obscure. The simple version is this: as sociobiology and evolutionary psychology attempt to biologize all of human nature and behavior, they fall into the trap of (or are in fact exploited by) reactionary and racist ideology. The counterargument is that the "truth" of science is apolitical, and that unless one biologizes the human, one misses fundamental truths about humanity and succumbs to the fallacy that human behavior is determined by cultural rather than biological factors. My point, however, will not be to criticize these phenomena—however uneasy they make me, I am in no position to do more than take advantage of the many and serious critiques and expositions of the work already in place—but to think of them as, precisely, another and later use of Darwin, one that is distinctly descended from the late Victorian uses we have just considered, and that remains extremely controversial on political and moral as well as scientific grounds. In the course of the debates over these issues, Darwin evokes all the critical battles that might have been waged directly over interpretations of his writing: biological determinism, reductionism, progress, social equality as opposed to biological and (therefore) cultural hierarchy. In attempting to explain all things as Darwin attempted to, in terms of natural law, the contest here has shifted from science and state to science and culture.

If in the course of this book I want to make the case for a use of Darwin that can in certain respects re-enchant our world, I need to confront some of the recent directions of evolutionary biology that seem to proceed even further along the road to disenchantment than the Victorians did, and that claim to go yet further in explaining all aspects of human behavior and achievement in terms of evolution by natural selection. As with Darwin's work itself, I want to look not so much at the coherence and ambition of the arguments (although they will also be important in the discussion), but at the *affect* of the language with which they are presented to a wider public, with the attitudes toward life itself that they imply. Clearly, Darwin is present (in some form or other) in the enterprises. He is invoked by virtually everyone. The mechanism of natural selection, seen with increasing mathematical sophistication, seems to have become the key to all mythologies.

One way to begin the discussion is with a comment by Richard Lewontin, one of the best-known antagonists in current controversies:

> There is a form of vulgar Darwinism, characteristic of the late nineteenth century and rejuvenated in the last ten years, which sees all aspects of the shape, function, and behavior of all organisms as having been molded in exquisite detail by natural selection—the greater survival and reproduction of those organisms whose traits make them "adapted" for the struggle for existence.[1]

Lewontin treats sociobiology and evolutionary psychology as "vulgar" Darwinism, a much too literal use of natural selection, aimed at establishing a direct connection between genetic inheritance and all aspects of human life. His antagonism is to what he sees as adaptationism, a view that he says badly underestimates the role of contingency in evolutionary development. (In this respect his arguments are similar to those of Stephen Jay

Gould, equally hostile to adaptationism, sociobiology, and evolutionary psychology.) Darwinian himself, Lewontin not only raises strong scientific objections to adaptationist readings of evolution but is troubled deeply by what he sees as the ideological work such strict adaptationism tends to do. There is a question—certainly one that would be raised by many evolutionary psychologists—about whether Lewontin's is an accurate representation of these sciences, but certainly it captures something of their direction and of the nature of the controversy that keeps Darwin alive in contemporary scientific debate. My interest here is not so much in sorting out adaptationist and anti-adaptationist positions as in simply registering the politically ambivalent ways in which Darwin's work is used at the moment, how important it is to current debates, and how ideologically and politically loaded it remains. Certainly, as I will be arguing, Lewontin is right in pointing out how Darwinian scientists today continue or pick up a strong tradition of late Victorian science.

I want, however, to further frame the discussion with the perspective of a modern Darwinian naturalist who predates sociobiology slightly but who was certainly himself a voice for enchantment, while remaining a true believer in natural selection. Loren Eiseley writes about nature as though it were alive with a significance not reducible to scientific analysis, as though it were full of mysteries that make it enchanted, despite the adequacy of natural selection as an explanation of the origin of species, despite the hard truth that the world is not disposed to please human beings nor animated by any of the gods that have conventionally occupied the imaginations of the West. His is certainly not the language of the sociobiologists to follow him. He occasionally expressed annoyance with Darwin (and the "harsh" late-Victorian propagandizers for science) for a too mechanistic emphasis on struggle, but Eiseley also recognized in Darwin a Romantic passion for nature that inflected all of his work, just the sort of Romantic attachment that later scholars like Robert Richards and Gillian Beer have also described.

As he considers the way science is actually practiced, Eiseley distinguishes two ways of being Darwinian. There is Eiseley's version of the kind of scientist Darwin himself was, who still thrills to the richness and intricacy of nature, and has "a controlled sense of wonder," and another, very different type who now certainly claims the Darwinian mantle. The Darwinian type of scientist feels

> wonder before the universal mystery, whether it hides in a snail's eye or within the light that impinges on that delicate organ. The second kind of observer is the extreme reductionist who is so busy stripping things apart that the tremendous mystery has been reduced to a trifle, to intangibles not worth troubling one's head about. . . . The *only* true reality becomes the chill void of ever streaming particles. (*Star Thrower*, 190)

So Lewontin's "vulgar Darwinians" are Eiseley's reductionists. Many thinkers who might be recognized among the second category would vigorously—angrily—deny that they practice the sort of reductionism that Eiseley condemns here, and would be appalled by his invocation of "mystery" in the midst of a subject that requires scientific attention. Eiseley himself, so given to celebration and exploration of that mystery of nature's processes and phenomena, might seem to many of them, to Richard Dawkins, for example, or to Daniel Dennett, a mere mystifier of nature's processes, an obstacle to serious scientific thinking, although Eiseley was a thoroughgoing evolutionist, an excellent historian of science and naturalist, and all too movingly aware of the heartlessness of nature's processes. He cannot be dismissed from the scientific scene as a simple sentimentalist, but his rhetoric often seems, certainly to reductionist ears, merely mystifying. Eiseley set himself against the work of demystification and disenchantment, against the notion of nature as a mechanism in which struggle is the dominant mode: "One can only assert that in science, as in religion, when one has destroyed human wonder and compassion, one has killed man, even if the man in question continues to go about his laboratory tasks" (197). Still, it is easy enough to recognize, at least at the furthest margins, the kinds of science Eiseley is talking about. He is right about Darwin's Romanticism. To represent *that* Darwin is one of the primary motives of this book.

I can't do so, however, by sentimentalizing Darwin, who surely does, in some ways, lie behind these disenchanting practices, as Eiseley knew and regretted. As early as July of 1838, in his "M" notebook, he was clearly plotting to disguise the degree to which he was committed to a materialist explanation for what would have been thought of as distinctively human characteristics: "avoid stating how far I believe in Materialism" (*Notebooks*, 532). But of course the notebook is full of materialism, including this delightful note: "It is an argument for materialism. that cold water brings on suddenly in head, a frame of mind, analogous to those feelings. which may be considered as truly spiritual" (524). In an entry shortly afterward he claims that "Now it is not a little remarkable that the fixed laws of nature should be <<universally>> thought to be the *will* of a superior being; whose natures can only be rudely traced out. When one sees this, one suspects that our will may <be> <<arise from>> as fixed laws of organization.—M. le Comte argues against all contrivance—it is what my views tend to" (537).

If affection drove his attention to his infant children, his notes on his first son, William, in infancy are still strong evidence of his unwavering naturalistic commitment to his argument for heredity. No one, he notes about the "moral sense," "has approached [the subject] from the side of natural history," and his determination to do so certainly foreshadows the project of sociobiology. "Any animal whatever," he claims, "endowed with well-marked social instincts, would inevitably acquire a moral sense or conscience" (71). Although, then, his published writings, his notebooks, and his life are full of evidences that a strictly literal interpretation of his words does not do justice to his sense of the way the world works, in a way Darwin is (or might be taken as) the patron saint of biological reductionism, and particularly of sociobiology and evolutionary psychology.

It is clear that contemporary sociobiology and evolutionary psychology—those most modern of Darwinian uses—are practices of the second category of Eiseley's scientists here. Let me quote a short passage by Steven Pinker to note precisely the sort of rhetoric that Eiseley means, and that gives sustenance to Weber's theory of disenchantment.

> Mental life can be explained in terms of information, computation, and feedback. Beliefs and memories are collections of information—like facts in a database, but residing in patterns of activity and structure in the brain. Thinking and planning are systematic transformations of these patterns, like the operation of a computer program. Wanting and trying are feedback loops, like the principle behind a thermostat: they receive information about the discrepancy between a goal and the current state of the world, and then they execute operations that tend to reduce the difference. The mind is connected to the world by the sense organs, which transduce physical energy into data structures in the brain, and by motor programs, by which the brain controls the muscles.[2]

I don't mean to fall back here in Romantic horror at these mechanical and computer metaphors. The *affective* work they do is self-evident. Although Pinker's book is clearly directed at explaining and popularizing the project of evolutionary psychology to a lay public, Pinker may have chosen the language simply because those are the metaphors that dominate the fields he works in. But surely he is too good a writer not to be exploiting their iconoclastic and demystifying effect. He, like naturalists of many stripes, but with perhaps a bit more aggression, wants to eliminate from our understanding of science, nature, and humanity any of the conventional assumptions about the god in the machine, about spirituality and nonnatural causes. (It's important to recognize that virtually everyone involved in the debates on these issues except the creationists themselves is committed to secular explanation if not to aggressively secular visions of the world.)[3] Although he writes extensively and interestingly in disclaiming the kind of extreme reductionism that Eiseley criticizes, Pinker *sounds* reductionist, even perhaps like Lewontin's "vulgar" Darwinian. There is, indeed, a moral urgency in his language (interestingly, because he complains of the moral/political agenda of opponents): "our culture can only be enriched by the discovery that our minds are composed of intricate neural circuits for thinking, feeling, and learning rather than blank slates, amorphous blobs, or inscrutable ghosts" (*Blank Slate*, 72).

The popularity of work like Pinker's suggests that there is something broadly attractive and even reassuring about these disenchanting moves, as if the culture as a whole—or some large, literate part of it—is desperately seeking clear scientific explanation and clarification of the muddy complications of human experience. The rhetoric, in any case, does another kind of mystifying work, even if its literal function is to undercut mystification, for it moves the reader from the common sense or folk apprehension of brain and mind to a sort of awed initiation into the complexities of computer science. It produces a kind of shock of astonishment when the flexibility and creativity of the human mind are transformed into a computer of the sort that any trained engineer might produce and that can be found now in profusion, along with television sets, in people's homes.

I am not addressing here the complications of Pinker's arguments, but only attending, with this quotation, to the *affect* of his rhetoric. It can give some indication of what it is that so repels Eiseley, but it also lays the groundwork for my discussion, in a later chapter, of Darwin's prose. It is interesting to think of it in relation to the work of the scientific naturalists, in particular W. K. Clifford, John Tyndall, and T. H. Huxley, who did so much in the late nineteenth century to disseminate Darwin's thought and to propagandize for the virtues of science as against the mystifications of religion. Like Pinker, they wrote with remarkable clarity about complex scientific issues, and like Pinker they demanded "scientific" explanation of human phenomena, flirting with a pure materialism (which they all, of course, denied), as Pinker flirts with reductionism. His arguments may not be those of "extreme" reductionism (I do not in fact think they are), but the rhetoric takes no pains to make the arguments *feel* anything but reductionist—its work is demystification, expelling the "ghost from the machine." This, I'm sure, would have been enough to make Eiseley claim that Pinker's science is dehumanizing.

Like the writing of the scientific naturalists, Pinker's prose is shot through with zest and even, one might say, Romantic energy. The unflinching scientist, telling it like it is without a nod to traditional ways of feeling, represents himself as a kind of culture hero. The voices of Victorian "freethought," like Pearson's, for example, regularly insist on the courage required to value truth above desire and to dispel old myths and mysteries. Indeed, William James amusingly complains that W. K. Clifford, making pronouncements of this sort, has "too much of robustious pathos in his voice" (*Will to Believe*, 8). So there can be (and is in many current forms of the argument) a latent, bleak romanticism even in the vision that confronts the world as a "chill void of streaming particles." But in making none of the sorts of "compromises" that Darwin made, sensitive to the sorts of disruption his materialistic reading of life and human experience produce, and almost certainly feeling something of that disruption himself (good Victorian gentleman that he was), Pinker prefers to agitate and push the implications of his computer metaphors as far as they will take him. Celebrating the personal heroism of accepting disenchantment as the true human condition, he is both Darwinian in his commitment to materialistic explanations and quite un-Darwinian. The *affect* of his writing has little to do with the wondrous nature of the phenomena science investigates, everything to do with the moral urgency of pursuing rational truth above any other forms of desire, and with the scientist's power to do just that.

With the preliminary act of disenchantment, reenacting the drama of the Victorians, who for the first time culturewide were encountering Darwinian news, there is rebellious enthusiasm in the rhetoric of "robustious pathos" that echoes again in Pinker's writing and in much of current popular writing that attempts to disseminate the argument for biologizing everything human. Steven Weinberg, a major physicist rather than an evolutionary biologist, brilliantly making the case for a "compromised reductionism," is in this respect characteristic: "The reductionist world-view *is* chilling and impersonal. It has to be accepted as it is, not because we like it, but because that is the way the world works" (52). The way the world works is scary, and Daniel Dennett too insists that the scariness has to be faced. Darwin's "dangerous idea," he says, explodes many of the great moral ideas of humanity. His book on Darwin, then, "is for those who agree that the only meaning of life worth caring about is one that can withstand our best efforts to examine it" (20–21).

Reaction to this kind of triumphant naturalism set in early among the Victorians, and the cultural triumph of its commitment to the moral and intellectual primacy of scientific rationality was at best partial. Among us, the partial triumph of this way of thinking has helped provoke a vast "counterculture" of New Ageisms and forms of spiritual silliness, as well as a frightening resurgence of Christian fundamentalisms. It is rather difficult to find the kind of serious response to the pervasively disenchanting projects of science among us that were frequent in the nineteenth and early twentieth century. William James cites Clifford's uncompromising view that "Belief is desecrated when given to unproved and unquestioned statements for the solace and private pleasure of the believer." "All this," James wrote, "strikes me as healthy." "Yet," he goes on to say, "if any one should thereupon assume that intellectual insight is what remains after wish and will and sentimental preference have taken wing, or that pure reason is what then settles our opinions, he would fly quite as directly in the teeth of the facts" (*Will to Believe*, 8).

Eiseley's critique of triumphant naturalism is the closest late-twentieth-century analogue to James's of Clifford that I know, and certainly he is right that the *kind* of argument Pinker makes—putting to the side the question of whether the argument is correct (after all, Weinberg agrees that such views are "chilling")—is part of that general ambition to total explanation that has made science, again in Weber's narrative, the instrument of modern disenchantment. Our current aggressively disenchanting post-Darwinian rhetoric does seem to echo with the Victorians' immediate post-Darwinian rhetoric; the moral/intellectual perspectives replay (without apparent self-consciousness—although Lewontin seems to have noticed) the battles of Victorian scientism at the end of the nineteenth century, some aspects of which I have intimated in my discussion of Pearson and Kidd. Heroic epistemological self-denial is the name of the game, as we are enjoined to have the courage to face the stern truth, the chill void, if that it must be.

The "dream of a final theory" now extends from Darwin to physics to human psychology and behavior. As Darwin claims at the end of the *Origin*, his work might lead to "open fields for far more important researches. Psychology will be based on a new foundation, that of the necessary acquirement of each mental power and capacity by gradation. Light will be thrown on the origin of man and his history" (488).[4] Roger Smith notes in his compendious history of the human sciences that "if any one name was attached to belief that there is or could be a natural science of human nature, it was his. The evidence that human beings have evolved from physical nature vindicated the conclusion that human nature and physical nature are understandable in the same terms."[5] Arguing that man is descended from some "lower form," Darwin can go on to assert that even "the moral sense" and "conscience" will show evidences of that descent (*Descent*, 1:71). Physiology will explain psychology: "violent gestures . . . increase rage." There is, Darwin argues, an "intimate relation . . . between almost all the emotions and their outward manifestations."[6] This is, of course, part of his overriding case that "there is no fundamental difference between man and the higher mammals in their mental faculties" (*Descent*, 1:35).

Moving then from biology to anthropology, even to ethics and aesthetics, Darwin follows the lines of a fundamentally reductionist argument, not in the extreme sense that Pinker and Dennett and virtually all other reductionists disavow, but in the sense of seeking the most fundamental explanatory theory to cover the broadest number of phenomena. Certainly, Darwin's arguments implied that at the root of all life—at its real historical source—there was one organism, and that organism could be understood not as a spiritual being but as a biological, material one. We are all brothers and sisters, and we are all significantly shaped by our organic past. This certainly sounds consistent with the contemporary efforts of sociobiology. Smith asserts that ardent Victorian evolutionists were excited by Darwin's powerfully documented argument "precisely because evolutionary theory unified human and natural science, and their descendants, such as the sociobiologists active in the 1970s and 1980s, shared this position" (455). Dennett finds Darwin's idea the most important in the history of science, "in a single stroke" unifying "the realm of life, meaning, and purpose with the realm of space and time, cause and effect, mechanism and physical law." The Victorian dream of a holistic science, a "consilience" of the sciences in an ultimate project of full explanation, moves from Darwin to E. O. Wilson.

Scientifically *and* ideologically, there is no escaping the battle over reductionism. Wilson, Dawkins, and Pinker—the most prominent (or widely published) figures in the public arguments for sociobiology and evolutionary psychology—all claim proudly to be reductionists, and all reject absolutely the idea of reductionism that is implicit in the quotations from Eiseley and Lewontin with which I began this discussion. Here is a point at which the margins between "scientific" and "political" arguments are obscured, but it seems pretty clear that the scientific revulsion from reductionism is fueled by a sense, more or less precisely articulated, that reductionism entails biological determinism, and biological determinism has a long history of collaboration with racism and fascism. It is worth pausing once again to consider some of the debates over reductionism, for its work and its possible political affiliations reflect directly upon Darwin's work, fairly or not.

Despite the arguments of David Sloan Wilson to which I earlier referred, and despite the fact that Elliott Sober has claimed that "If there is now a received view among philosophers of mind and philosophers of biology about reductionism, it is that reductionism is mistaken,"[7] modern applications of Darwin tend increasingly, like E. O. Wilson's, to be strongly reductionist. Reductionism is, for example, the intellectual cornerstone of Daniel Dennett's *Darwin's Dangerous Idea*, in which he claims that Darwin's dangerous idea "is reductionism incarnate."[8] But Dennett wants to make a distinction among reductionisms. While he too indulges in the rhetoric of contempt for antireductionists and claims that extreme reductionism is a mere straw man, he does reject one kind of reductionism, the kind he calls "greedy." Greedy reductionism, he says, "might lead us to deny the existence of real levels, real complexities, real phenomena." So the problem, according to Dennett, is not reductionism but bad science, moving far too quickly to large inferences of connection among disparate behaviors possibly explicable in many other ways, and underestimating the way in which the nature of higher-level organisms and groups cannot be reduced to the functioning of their smallest components.

Dennett's response to critiques of reductionism parallel most of the others I have encountered. There is good reductionism and bad reductionism, and it's the bad kind that allows all those nasty connections because, of course, they are bad science as well. Richard Dawkins, self-proclaimed reductionist, provides an excellent, if characteristically aggressive, definition of both kinds of reductionism:

> The non-existent reductionist—the sort that everybody is against, but who exists only in their imaginations—tries to explain complicated things *directly* in terms of the *smallest* parts, even in some extreme versions of the myth, as the *sum* of the parts! . . . it goes without saying—though the mythical, baby-eating reductionist is reputed to deny this—that the kinds of explanations which are suitable at high levels in the hierarchy are quite different from the kinds of explanations which are suitable at lower levels. . . . Reductionism . . . is just another name for an honest desire to understand how things work.[9]

Here Dawkins attempts to dispel one of the major complaints about reductionism, that it attempts to explain all levels of the hierarchy of organism and components in the same way, so that (as he is often probably misconstrued) humans can be explained entirely in terms of the "selfish genes" that generate them. But it turns out on Dawkins's and Pinker's accounting that there *are* no extreme reductionists. The extreme, "preposterous" reductionists who "want to abandon the principles, theories, vocabulary, laws of the higher-level science, in favor of the lower-level terms" (Dennett, 81)—there are no such nuts. They are "nonexistent," argues Richard Dawkins.

John Dupré has his doubts about this and attacks the reductionism of evolutionary psychology, clearly with Dennett in mind:

> Despite lip service to the importance of interaction with a variable environment, strategies of investigation that start with the influence of the environment are seen as worthless and the only properly scientific way forward is investigation of the intrinsic tendency of the organism to emit behaviour of various sorts. The environment ends up as little more than a trigger that determines a selection from among the range of internally generated behaviours.[10]

But Dupré's point goes beyond this, for he wants to claim that reductionism, in whatever form it might take, does nothing one way or the other to prove the validity of evolutionary psychology. Pinker and Dennett, in a crusade to expel the ghost from the machine, to deny the blank-slate theory of development, to make the study of human nature a science, want to emphasize the purely genetic and Darwinian processes that produce evolutionary development. All evolutionary psychologists want to insist that there is such a thing as a universal human nature, constructed everywhere out of the workings of natural selection.

It is difficult to argue against the idea that the biological has something to do with who we are. "No doubt," Dupré says, "the suggestion that the human sciences should depend more heavily on biology than has previously been allowed will appeal to a reductionist sensibility . . . but not even the most rabid evolutionist denies that behaviour varies in significant ways according to social context, and even the most die-hard culturalist agrees that there are organisms of a specific if highly variable kind" (75). But what needs to be done is to work out precisely the ways "in which our biological nature interacts with countless aspects of our environment to produce behaviour." Reductionist theory cannot resolve these difficult empirical questions.

The association of evolutionary psychology with a stern reductionist program, along with its attempt to minimize the importance of environment in the interplay between biology and culture that everyone concedes, keeps evolutionary psychology controversial and spurs the continuing association of its program with a long history of biologism that has had awful political consequences. It is certainly true that there are no inevitable connections between general scientific propositions and particular political ones, but Pinker is probably at least partially correct when he argues so vehemently that attacks on the new sciences are a consequence of the culturalists' assumption that biologizing human nature opens the way to such horrendous movements as eugenics and the racist science that runs from Victorian craniology to Jensen's IQ studies, to *The Bell Curve*. Juxtaposing the biologizing of human nature to its opposite (Pinker's bête noir), contructivism, Pinker argues that "A denial of human nature, no less than an emphasis on it, can be warped to serve harmful ends. We should expose whatever ends are harmful and whatever ideas are false, and not confuse the two" (*How the Mind Works*, 48). This seems to me completely right. But that there are contingent connections almost all the time, and that biologizing has a nasty history, the record confirms. These associations are real enough, and yet the actual politics of people like Pinker, Dawkins, and Dennett are hardly either racist or conservative. Whether or not reductionism is one's method, the political and ideological direction one takes will still have to depend on the usual things: what questions are asked? what data are used? how they are to be interpreted? There is an awful lot of culture in all of these aspects of investigation.

The problem with so much of the polemic circling around reductionism and its ideological propensities is that while everyone of any intelligence involved in the debate concedes that both biology and culture are at work in human behavior and psychology, straw men take most of the fire: the SSSM, the standard social science model, sneeringly derided by Steven Pinker; culturalists who insist that we are mere "blank slates" when we enter the world; and "biological determinists" like E. O. Wilson, Pinker, Dawkins, and Dennett, none of whom confess to biological determinism (Wilson is perhaps another matter, and I'll be looking at his arguments in a moment). In fact, Dupré's phrase, "reductionist sensibility" suggests a lot about what is at stake in this controversy. It is not so much the intellectual coherence or even the heuristic power of the reductionist stance; it is rather a way of being, thinking, and feeling—as we shall see most particularly in Wilson. Yes, culture plays a role somewhere, but reductionism is a deeply satisfying perspective to take on the world, and of course it tends to be a radically disenchanting one.

The book that, I believe, most passionately, intensely, and perhaps most unscientifically makes the case for reductionism, the temperament it represents, and the absolute ambitions of sociobiology and evolutionary psychology is E. O. Wilson's *Consilience*. To the disenchanting energies of the scientizing and biologizing bent of the new sciences, Wilson responds by seeking to refill the world with enchantment, but unlike James one hundred years ago, without yielding an inch to the pulls of desire that official science rejects as corrupting of objectivity. Wilson's book is a great "apology," in the sense of Newman's *Apologia pro vita sua*, providing a historically based defense of the efforts to biologize the human, and Wilson is himself a deeply engaged and effective conservationist and propagandizer for biological diversity. His work has a deep Victorian and Romantic tinge about it, and in its romanticism and wonder at the beauty, diversity, and sublimity of the natural world is also Darwinian.

But for Wilson, the Romance comes out of reductionism, and of all the figures to whom I have alluded thus far, he is the most insistently reductionist, the figure who is most committed to explaining higher organisms in terms of lower, who seems most ready to make the claim, as Lewontin and others describe it, that "the compositional units of a whole are ontologically prior to the whole that the units comprise."[11] I want, then, in concluding this excursion into the ways Darwin now gets used in "science" and in social ideology, to look at Wilson's most recent extensive restatement and development of the sociobiological program, *Consilience*, as it spins once again the Darwinian inheritance—and in politically ambivalent ways.

Consilience is important here in part because it is so clearly written with a moral and even spiritual goal, even as it pursues with self-conscious intensity the great holistic enterprise of the Enlightenment, an enterprise that would unify all knowledge. Although his end is rather similar to Pinker's, Wilson does not exploit demystifying language in the way Pinker does, although he certainly does not abjure the technical. His description of the brain, for example, is metaphorical and anecdotal:

> Its fluffy mass is an intricately wired system of about a hundred billion nerve cells, each a few millionths of a meter wide and connected to other nerve cells by hundreds of thousands of endings. If we could shrink ourselves to the size of a bacterium and explore the brain's interior on foot, as philosophers since Leibniz in 1713 have imagined doing, we might eventually succeed in mapping all the nerve cells and tracking all the electrical circuits. But we would never thereby understand the whole. (*Consilience*, 97)

This is gentle and deliberately "enchanting" prose, allowing the computer circuitry to emerge but encasing it in a fully human perspective and narrative. The rhetorical object here, surely, is not the disenchanting shock of Pinker but an almost romantic perspective on a mathematical sublime. So here we see that the most radically reductionist program of sociobiology is rendered in a language that attempts to re-enchant.

Wilson approaches his subject from a background very different from Pinker's. The rather autobiographical first chapter is, quite deliberately, called "The Ionian Enchantment," a great scientific counter-enchantment, "the dream of unified learning" (3). This is the enchantment with which Wilson seeks to replace the enchantment that he sees having grown out of prescientific ignorance. It means, says Wilson, "a belief in the unity of the sciences—a conviction, far deeper than a mere working proposition, that the world is orderly and can be explained by a small number of natural laws" (4). It is the dream of nineteenth-century Positivism, which imagined itself as a replacement for the old and now incredible religion, and which, in Auguste Comte's hands, became an actual, if not very successful, church. There is a quasi-religious aspect to Wilson's project, and he would not deny it: "Could Holy Writ be just the first literate attempt to explain the universe and make ourselves significant within it? Perhaps science is a continuation on new and better-tested ground to attain the same end. If so, then in that sense science is religion liberated and writ large" (6).

There can be no question that Wilson's enterprise is partly shaped by his falling away from religion and his almost instinctive need to replace it (another very Victorian phenomenon). While being scientific is the ideal condition to which Wilson aspires and which he wants his theories to meet, there is a strong ideological tug at work in his undertaking, as is clear from the autobiographical first chapter and the third chapter, which offers a history of modern knowledge. Ullika Segerstråle makes this point, based on her conversations with Wilson:

> He wanted to make sure that there could not exist a separate realm of meaning and ethics which would allow the theologians to impose arbitrary moral codes that would lead to unnecessary human suffering. He believed that there must exist a natural ethics for humans and was on the lookout for it. For Wilson, any new scientific knowledge which could allow human beings increased control over their lives would take power away from the theologians who want to run other people's lives. (38–39)

"Natural ethics" is a point of contention in an extended philosophical debate, going back well before Mill's "Nature," and the belief in "natural ethics," the idea that the right ethical system is out there waiting to be discovered (by scientists, of course), is and must be a faith; it cannot be empirically proven. One might of course "empirically" demonstrate that certain human moral codes derive from the necessities of natural selection, but even that wouldn't establish that the codes were "true" or "right." On the lookout for such a "natural" ethics, from which there will be no escape into irrationality and superstition, Wilson claims that "The greatest enterprise of the mind has always been and always will be the attempted linkage of the sciences and humanities" (8). So Wilson begins a book that will in fact attempt to make the linkage, or at least to prepare the way for it, a book that will be sermon, exhortation, analysis, and description, all at the same time. The linkage depends on what is surely a metaphysical assumption precisely about the ultimate orderliness of the world, and therefore of the possibility of reductionism. There is something admirable about the ambition and even the ideal, but here the devil is certainly in the details, as the repeated historic failures of the ideal suggest. Dupré has argued extensively against the metaphysical notion of the "unity of the sciences," insisting rather on "the disorder of things."[12] The battle over these issues, in which—not surprisingly for a literary critic— I find Dupré far more convincing than Wilson and his kind, is too complicated for any but an extensive philosophical/scientific investigation.

The battle makes clear the fact that the question is ultimately not a scientific one at all but a metaphysical one. Wilson's book (and to a certain extent he acknowledges this in passages I have already quoted) depends on his reductionist "sensibility," on his commitment to a set of ideas not provable, or provable only if a totally unifying science in fact comes into being. The metaphysics has its heuristic value, then, but whether the world is ordered or disordered remains (and my metaphysical bet is that it will remain) moot.

Wilson's enterprise is striking both in its ambition and in what might be called its naïveté. Segerstråle, in a work that attempts to analyze sociologically the intense responses to sociobiology (and which, on the whole, comes down against the critics, like Lewontin, for having in effect blindsided Wilson and for pushing political agenda too hard), talks of Wilson's political naïveté before the publication of *Sociobiology*. "In 1975 I was a political naïf," Wilson has written, describing how unprepared he was for the Marxist attacks on what looked to them like biological determinism.[13] And although Wilson was ready to argue against the political use of sociobiology, even at the time of publication, he seems in *Consilience*, written after the entire experience and after mulling quite seriously on all of the issues, to have retained the naïveté about the possible political implications of his unifying and reductionist program. The great, enchanting goal of the unification of knowledge clearly remains for Wilson *the* great moral good that would, in the end, even lead to social and cultural conditions in which racism, sexism, and the other ills that society is heir to would be minimized. There could be no escape from a "natural ethics."

To a literary reader, the moves of *Consilience* toward its unifying argument are breathtakingly ambitious and self-confident. In the third chapter, Wilson offers a narrative of the history of science and knowledge since the Enlightenment: "The faith of Enlightenment thinkers in science was justified," he tells us (45), and he goes on to provide the conventional hagiography of great figures and passes the conventional judgments on them. In a moment in intellectual history when just such hagiography is being written out of the history of science, to be replaced with a rich contextualism, Wilson quite comfortably tells the old story, one that Victorians, particularly the scientific naturalists, told over and over again as they pushed for the new authority of science. Wilson does it with a similar passion, a passion that it is impossible to miss. That passion is for the ideal of the order of nature, and those in the history of thought who have done most to further, by Wilson's lights, the recognition of that order are the heroes of his history. In a way, Wilson's history of Western thought plays out the "romantic" aspect of the late nineteenth century to which I earlier alluded, the concept of the heroic (and able) intellectual adventurer, seeking the truth at any cost. No doubt, this is Wilson's self-genealogy, for *Consilience* is a book, following upon *Sociobiology*, that offers itself as one further, large step toward the vision of order.

The progressive narrative Wilson tells is interrupted by the emergence of Romanticism. Despite his passion for reason, or perhaps because of his *passion* for reason, Wilson understands the possibility of disenchantment in a secularized world from which the mysteries of earlier mythic and metaphysical fantasies are barred. Recognizing the problem of an affectless rationality, and aware too that "a science-driven society risks upsetting the natural order of the world set in place by God or, if you prefer, by billions of years of evolution" (34), Wilson provides a relatively sympathetic description of the Romantic mistake, and mistake, of course, is what he takes it to be: "With Wordsworth's 'breathings for incommunicable powers,' the eyes close, the mind soars, the inverse square distance law of gravity falls away" (35). We hear of Rousseau's "stunningly inaccurate" anthropology, of Goethe being "easily forgiven"—but that, of course, is because "he had a noble purpose, no less than the coupling of the soul of the humanities to the engine of science" (36). Wilson concedes that the many admirable Romantic mistakes gave way to some great science, but, he is convinced, Romanticism frightened scientists away from the examination of mental life, "yielding to philosophers and poets another century of free play" (37).

"Free play." It is a phrase of condescension, patting the ignorant baby on the head. This is one of those breathtaking moments Wilson's history of the sciences produces. It is a stunning phrase, implying both Wilson's great ambition, his metaphysical faith in system and order, and his ultimate contempt for any intellectual project that does not commit itself with scientific rigor (this is not the place to discuss whether there is such a thing as a unified, definable scientific method) to the study of the natural world, which includes, of course, human mind and behavior. As one revels in feeling, one must never take one's eye off the "inverse square distance law of gravity." There can be no free play in any serious encounter with the world that is not meant simply to evoke emotion. The Ionian enchantment seems aimed at an ultimate disenchantment.

Romanticism gave art and the humanities one more century of "free play," but *Consilience* announces that play time is over. Not, however, before Wilson lashes out, with remarkable fury for so gentle and reasonable a man, against "postmodernism," or "poststructuralism" or "deconstruction," all of which are of course committed to disorder and nonknowing. Wilson claims to have read Derrida and to know, as far as Derrida is intelligible, what he and other postmodernists are saying. At last, he concedes that they have one virtue: "Their ideas are like sparks from firework explosions that travel away in all directions, devoid of following energy, soon to wink out in the dimensionless dark. Yet a few will endure long enough to cast light on unexpected subjects" (44).

The problem for me with this all too conventional and superficial reading of intellectual history and the movements of modern thought is not really the narrative itself. While it is familiar, I have found various Victorian renderings of it (without, of course, the finale on postmodernism), Huxley's, for example, that seem both more rigorously tied to historic fact and, as prose, marvelous, sometimes inspiring, reading. If it is a little more difficult now to believe in the progress implicit in such narratives (and to read out of them the ideological forces that also impel them), the story itself has the quality of a scientific pilgrim's progress, beset with great temptations and demanding great courage. My own orientation being (at least) as secular as Wilson's, and my commitment to the improvement of knowledge as against superstition being as passionate, I can't object much to the narrative except that it is a very old story and has lost some of its earlier freshness and zap. It seems to be historically uninformed about its own predecessors, and thus even less likely to succeed. Moreover, it implies that the singular shape of Wilson's version of the narrative is the only possible shape, that unless one believes implicitly that the world is (secretly) totally ordered, and commits oneself absolutely to the pursuit of the full unification of knowledge that will demonstrate and make use of that order, one is merely a Romantic (that is, engaged in free play) or now—much worse, of course—a postmodernist.

This sort of cast-iron fitting inevitably provokes hostility, and not merely from "postmodernists." It is rather like the absolute dualism proposed by Pinker, when he distinguishes reductionists from constructionists, people who allegedly believe that there are no biological components to human behavior. That dualism is, like Wilson's, far too absolute. Again, a belief in the unity of the sciences is not a *condition* of good science; a belief in a reductionist understanding of the relations of the sciences to each other is not a *condition* of good science; a strong recognition of the ways in which such thought threatens to turn current conditions into natural and unchangeable ones is not, conversely, equivalent to a commitment to the contemptuously imagined SSSM version of knowledge. Pleading for consilience, Wilson would probably have done better to be less naïve about alternative models of science and rational doubts, scientific and ideological, concerning the movement toward consilience that he seeks.

The political naïveté of the Wilson who was unprepared for the attacks on *Sociobiology* from scientists on the left is matched by the new, merely apparent sophistication of a Wilson who has weathered that storm, has largely shifted the blame to a know-nothing postmodernism, and who fails once again to see how easy it might be to use the unification narrative in enterprises that are socially very dangerous. Wilson asks, "why should the social sciences and the humanities be impervious to consilience with the natural sciences? And how can they fail to benefit from that alliance?" (11). Ingenuous or disingenuous, the question suggests real incomprehension, even now, about what might be at stake in accepting a biologically reductive theory as a way of interpreting cultural phenomena. The narrative is partly written to let readers understand that the author knows about such dangers. Hence the comment: "Science given too much authority risks conversion into a self-destroying impiety" (34). But Wilson's effective answer to this is to deny that he is asking for "too much authority" for science. He wants rather "consilience," a bringing together of the humanities, social sciences, and natural sciences, in which reason will reveal order, with a soul.

The word needs consideration, but it is important here to remind ourselves of how Darwin fits into this narrative, although he does not figure much in Wilson's short history of post-Enlightenment thought, except as he "scandalized" Agassiz. But natural selection is the heart of Wilson's move to consilience, and, as we have already seen, in his vision of "open fields," Darwin seems himself to have been committed to the kind of narrative of knowledge that Wilson offers. He too tried with large speculation to bring a vast range of phenomena under the rule of law, and his revulsion from the interpretation of his theory as somehow invoking chance is well known. But Darwin, is both similar to and different from Wilson, and of course, for most of this book I will be wanting to talk about the ways in which he is different.

For Darwin, too, "reason" would seem to have been the dominant intellectual value as it is for his modern successors. But as one reads the *Origin*, one finds that reason does astonishing imaginative work, in effect, replaces the imagination as the faculty most capable of extending beyond the constraints of ordinary consciousness. In one striking passage, we can detect the reversal:

> He who will go thus far, if he find on finishing this treatise that large bodies of facts, otherwise inexplicable, can be explained by the theory of descent, ought not to hesitate to go further, and to admit that a structure even as perfect as the eye of an eagle might be formed by natural selection, although in this case he does not know any of the transitional grades. His reason ought to conquer his imagination; though I have felt the difficulty far too keenly to be surprised at any degree of hesitation in extending the principle of natural selection to such startling lengths. (188)

There is a kind of somersault here, and reason becomes the risk-taking, imaginative force, while "imagination," in a characteristic Darwinian inversion, holds back from extending over those "large bodies of fact" and incorporating the most extravagant and difficult facts of all. Reason is the right word for the scientist, but in Darwin's hands, reason outstrips the imagination in its

power to break from conventional conceptions; and that is because the world Darwin describes is so uncannily organized, so
Romantically, astonishingly, complex and diverse. It is a world
for which the nonscientist would normally invoke the imagination. But it is fair to suggest that Darwin would have hesitated
at "free play" in the sense Wilson has used and denigrated. The
fullest freedom comes from the disciplined and, indeed, self-
denying energies of rationality. In this sense, Darwin is rather
Wilsonian.

Darwin, however, is only part of the story. Wilson claims that
the project of sociobiology, while at an extremely early stage, has
met the first problem, "altruism," with successful research and
is moving forward as a full-blown science. He admits, however,
that "human behavior genetics is an infant field of study and
still vulnerable to ideologues," but that in the "estimation of her-
itability" it has become "an advanced scientific discipline" (154).
Consilience in effect lays out a program for sociobiology, one that
will bring together all the human sciences and all the humanities
with the biological work built out of evolutionary biology to arrive at last (and Wilson is utterly confident of this) at a fully satisfying science of the human. It was Comte's vision too, and
Comte is invoked warmly by Wilson because he too sought
(what he did not call an ultimate) "consilience," science building
upon science until a full science of the human would be constructed on the hierarchical generalizations from the lowest sciences to the highest. "Consilience," some hundred and fifty years
later, sounds rather like Comte's formulation, familiar from our
recent discussion of reductionism; "explanation of facts is simply
the establishment of a connection between single phenomena
and some general facts, the number of which continually dimin-
ishes with the progress of science."[14] Unlike Comte, however,
Wilson concedes that the ideal of consilience is not yet science:
"it is a metaphysical world view, and a minority one at that,"
whose work at the moment is primarily heuristic, "the prospect
of intellectual adventure and . . . the value of understanding the
human condition with a higher degree of certainty" (9).

Wilson quotes Whewell, in his *Novum Organum Renovatum*, to explain what he means by "consilience":

> The Consilience of Inductions *takes place when an Induction, obtained from one class of facts, coincides with an Induction, obtained from another different class. This Consilience is a test of the truth of the Theory in which it occurs.*[15]

This does not seem to me to coincide precisely with Wilson's use of it. Laura Dassow Walls has suggested that Wilson's adoption of Whewell's useful term is "ironic," because Wilson "quickly moves on to redefine the term in his own way not as truths that have 'jumped together' but as the ingestion of one truth by another" through a process of reduction whereby "the laws and principles of each level of organization" are folded into those of "more general, hence more fundamental levels." The difference between Wilson and Whewell, she notes, is that Whewell's ultimate goal is not "one great fundamental truth but a network of total connectivity, which could be travelled in a dizzying array of directions. By contrast, Wilson's approach 'reduces' literature to an epiphenomenon of the laws of physics, losing Whewell's sense of emergence into ever 'higher' levels in the name of distillation into levels that are ever 'lower.'"[16]

We are back once again to Wilson's metaphysical commitment to reductionism. Whewell's theory is not necessarily reductionist. He simply argues that when an induction works in one area, it can be confirmed because it works in another area as well. The more such happy confirmations there are, of course, the richer the general induction. But Walls is right that the hierarchical understanding of laws absorbing lower-level laws across a Comtean universe of subjects is not part of what Whewell meant by "consilience."[17]

The details of the grand utopian narrative Wilson wants to write into the consilient future depend on our understanding how the brain works. In fact, Wilson argues, the logical-positivist program, "the most valiant concerted effort ever mounted by modern philosophers" (64), only failed, or fell short, because it did not have that understanding. Such stunning and unqualified observations make the book seem simultaneously naïve and so-phisticated, a trustworthy guide to the possibilities of a new sci-ence and at the same time a fantasy aimed at inspiring readers with the thrill of the possibilities of scientific achievement. Let me quote here an extensive passage which will register, I be-lieve, the way in which Wilson wants to take science beyond the disenchantment of full explanation that Weber noted to a new, Romantic enchantment (in effect, very like the nineteenth-century Positivism that Wilson admires). The passage will—in the very texture of the prose—carry out that "consilience" between sci-ence and the humanities and bring us to the promised land:

> If the exact biological process of concept formation can be defined, we might devise superior methods of inquiry into both the brain and the world outside it. As a consequence we could expect to tighten the connectedness between the events and laws of nature and the physical basis of human thought pro-cesses. Might it be possible then to take the final step and devise an unassail-able definition of objective truth? Perhaps no. The very idea is risky. It smells of absolutism, the dangerous Medusa of science and the humanities alike. Its premature acceptance is likely to be more paralysing than its denial. But should we then be prepared to give up? Never! Better to steer by a lodestar than to drift across a meaningless sea. I think we will know if we come close to the goal of our predecessors, even if unattainable. Its glow will be caught in the elegance and beauty and power of our shared ideas and in the best spirit of philosophical pragmatism, the wisdom of conduct. (65)

There is no denying the intensity of Wilson's passion for order and meaning and for the one vehicle by which these can be found, science. It is real and heartfelt. The world, threatened by "free play" or by those who think that the ultimate order of nature is not knowable or is not, ontologically, out there, is un-deniably enchanted for Wilson. This is the language of religion, incorporating the narrative of the brave intellectual quest, at whatever cost, for truth. "Never!" The "lodestar" is right out of Carlyle. This is prose that manifests the same desperation that critics have noted in Carlyle's prose for over a century; in its enor-mous aspiration and rhetorical fervor it aims at becoming a prose for the enchanted. But then it is quite a long way from the empirical rigor that Wilson manifests in his own scientific work. The "Never!" is in contest with the threat of overweening arrogance (or of, perhaps? political and social mistakes). But this enchantment of ultimate order, as it echoes the ambition of nineteenth-century thinkers to find an adequate replacement for the "lost Christian totality," seems just as likely to fail culturally.

It is perhaps unreasonable, in detecting these Victorian echoes, to recall the radical failure of Comte's Positivist religion in the nineteenth century. The thrill of the supreme rational empiricist in the faith that new knowledge will finally remake the world aroused few beating hearts among the Victorians, at a moment when a crisis of belief was purportedly widespread. But that it was a noble (if naïve) dream is not for me in question. Here, rather, I want to attend to the curious fact that the enterprise that evokes Pinker's iconoclastic computer analogies entails for Wilson a language of Romantic aspiration. Tennyson's Ulysses echoes here: "to strive, to seek, to find, and not to yield." In his own prose and in his own imagination of what's possible and what is necessary, Wilson incorporates the Romanticism that his narrative of Western intellectual history treated as an understandable mistake to reenforce the grand ideal of overarching Reason. Romantic feeling gives power to scientific knowledge. This is language that evokes not argument but prayer.

As with Darwin, one can move through Wilson's prose and pick out moment after moment in which the sense of beauty and awe accompanies hard-headed "scientific" argument. So Wilson discusses discoveries about living cells that scientists are now making: "The machine the biologists have opened up is a creation of riveting beauty" (91), and we move on toward the mathematical sublime through the nucleic acid codes, each "typical vertebrate animal" comprising 50,000 to 100,000 genes, each gene "a string of 2,000 to 3,000 base pairs." With the equipment of his enormous faith, his genuine excitement and belief that at last we have moved to a point in intellectual history when the bridge between body and spirit can be crossed, when we can produce a true science of the human ("as clear as the road to Charing Cross," said James Mill two hundred years ago), Wilson becomes a voice for progress:

> It is, I must acknowledge, unfashionable in academic circles nowadays to speak of evolutionary progress. *All the more reason to do so.* In fact, the dilemma that has consumed so much ink can be evaporated with a simple semantic distinction. If we mean by progress the advance toward a preset goal, such as that composed by intention in the human mind, then evolution by natural selection, which has no preset goals, is not progress. But if we mean the production through time of increasingly complex and controlling organisms and societies, in at least some lines of descent, with regression always a possibility, then evolutionary progress is an obvious reality. In this second sense, the human attainment of high intelligence and culture ranks as the last of the four great steps in the overall history of life. They followed one upon the other at roughly one-billion-year intervals. The first was the beginning of life itself, in the form of simple bacteriumlike organisms. Then came the origin of the complex eukaryotic cell through the assembly of the nucleus and other membrane-enclosed organelles into a tightly organized unit. With the eukaryotic building block available, the next advance was the origin of large, multicellular animals such as crustaceans and mollusks, whose movements were guided by sense organs and central nervous systems. Finally, to the grief of most preexisting life forms, came humanity. (98)

The touch of *ressentiment* in Wilson's italics makes clear how self-consciously *Consilience* challenges both the criticism Wilson encountered after the publication of *Sociobiology* and the intellectual movements that have insisted on indeterminacy, denied the possibility of objectivity, dismissed positivism as a philosophical movement, and raised doubt about the ultimate orderliness of the world. The book does not attend directly to the critiques of culture that require of all intellectual enterprise a sense of historical and social context (although Wilson does tell Foucault not to be so depressed [43]). If teleology is gone, "progress" gets smuggled back in through the notion of rational advance. As Robert Young has noted, what spiritual succor evolution took away, "It gave back in the doctrine of material and social and spiritual progress" (16). It is this succor that *Consilience* is devised to provide.

The affirmation of progress reflects the book's entire tone, an almost absolute certainty not only that science is the way but that the problems of nonscientific disciplines are curable with an injection of science. Questions that have literally evoked centuries of debate can be resolved very simply. Just clarify your terms a bit and then look at the facts. The fact is that there *has* been progress, in a straight-line, Herbert-Spencerian way, from the first living organisms to the production of human intelligence and culture. What is striking about Wilson's plunge into these issues is that this deeply learned and intellectually passionate man is utterly unintimidated by the long scholarly traditions of other disciplines. After all, he is seeking "consilience," which will bring together, ultimately, all disciplines; nor is he even now deeply reflective about the often shabby social work to which ideas like this have been put. However Darwinian and aware of the work contingency does in evolution, he seems to forget contingency when he writes of the progress of ideas. The thrill of the ideal world he imagines emerges against the "regression" that is "always a possibility," because over the long haul, evolution has in fact produced the crown of life, "high intelligence and culture." The echoes of Herbert Spencer in Wilson's arguments should not make for guilt by association.[18] Wilson clearly does not invoke progress on the smaller scale that for the Victorians often ended in racism, a vision of increasing complexity through anthropoids and dark-skinned peoples up the racial scale to the modern European. It is precisely the Romantic naïveté of his sophistication that is so striking.

As Wilson and Pinker rather bitterly note, distrust provoked by this kind of history led to what they take as an irrationally hostile rejection of sociobiology. Pinker—I infer, far to the left of Wilson politically—concedes that "many hereditarian movements *have* been right-wing and bad, such as eugenics, forced sterilization, genocide, discrimination along racial, ethnic, and sexual lines, and the justification of economic and social castes" (47). But, he notes, constructivist arguments that people and institutions can be changed by "re-engineering society" have themselves produced some pretty horrible consequences.

Wilson knows and speaks clearly about the way human behavior genetics, for example, is "still vulnerable to ideologues" (154). It won't be, however, once the science is more advanced. And in the nature/nurture battles that I have already noted, Wilson recognizes the interplay of genes and culture: "genes prescribe epigenetic rule," "culture helps determine" which genes survive, new genes alter epigenetic rules, and these new rules "change the direction and effectiveness of the channels of cultural acquisition." The contentious question—and indeed it is contentious—is the length of "the genetic leash" (157). How tightly are the epigenetic rules constrained by the genes? How much are the details of the culture and individual behavior constrained by the genes? Wilson can return to Darwinian investigations (as, for instance, the expression of emotion in man and animals) to note that there are universal signs for certain affects, like pulling down the corners of the mouth to expose the teeth to express contempt.

The recognizable and admirable positivist effort to turn the study of human activities into a science—an effort that the British nineteenth century saw in its importation of Comte's ideas, in the work of the Mills (James and John Stuart), and in the work of Darwin himself—has descended these days to evolutionary psychology. Darwin never expected that his explorations of human origins would lead to resolutions of the intricate complex of political, social, and cultural problems that sociobiology claims to address. Yet there seem no questions that Wilson doesn't imagine sociobiology helping with, and his sociobiological followers follow him in this. A textbook of evolutionary psychology proclaims, "only within the past few decades have we acquired the conceptual tools to synthesize our understanding of the human mind under one unifying theoretical framework—that of evolutionary psychology" (Buss, 4). Taking us from sperm cells to the details of modern behavior, the evolutionary psychology that makes these unifying claims will go on to provide formulas that predict how women read classified ads, how men and women think about whom their infants resemble, how much rape there will be and how impossible it is to purge it from human society, who will murder whom, and so on. For the unconverted, much of this seems like greedy reductionism.

Wilson imagines, as an example of the sort of unification that interests him, a problem of environmental policy. He indicates that now the disciplines of ethics, social science, and biology, all required in dealing with the problem and all closely connected, are nevertheless treated by academia as separate, with their own practitioners and methods. Here the result is "confusion." But, he argues, to solve problems in the real world we must be able to consider the relationships, focusing in on real-world problems, where "fundamental analysis is most needed," so that we might travel "from the recognition of environmental problems and the need for soundly based policy, to the selection of solutions based on moral reasoning; to the biological foundations of that reasoning; to a grasp of social institutions as the products of biology, environment and history. And thence back to environmental policy" (10). As an injunction to interdisciplinarity, this seems to me unexceptionable. As a suggestion that somehow as we move back to a recognition of the biological foundations of all cultural phenomena we will manage to avoid "confusion," it seems highly dubious. Where exactly does the unification Wilson seeks exist? Not much below the surface of Wilson's inclusion of ethics, moral reasoning, and the social sciences in the conversations is the reductionist view that Dupré decries, which ultimately assumes the priority of "the intrinsic tendency of the organism."

In any case, *Consilience* is not a scientific book; it is the intelligent and learned expression of a faith that if science succeeds (and the gaps are being filled in, Wilson constantly reminds us), all disciplines, biologized, will be more precise and more effective. In *Sociobiology* there was the implication that philosophy will be displaced by biology, for the biologist, who is concerned with questions of physiology and evolutionary history, realizes that self-knowledge is "constrained and shaped by the emotional control centers in the hypothalamus and limbic systems of the brain" (3). The implication becomes explicit in *Consilience*. We will, with biology, achieve better economic theory, better social theory, better aesthetic theory, and the ultimate replacement of religious transcendentalism with scientific empiricism:

> Religion will possess strength to the extent that it codifies and puts into enduring, poetic form the highest values of humanity consistent with empirical knowledge. That is the only way to provide compelling moral leadership. Blind faith, no matter how passionately expressed, will not suffice. Science for its part will test relentlessly every assumption about the human condition and in time uncover the bedrock of the moral and religious sentiments. The ultimate result of the competition between the two world views, I believe, will be the secularization of the human epic and of religion itself. (265)

Although this may seem, to secularists, a consummation devoutly to be wished, it is breathtaking in its imperialist takeover. Religion will become science, jazzed up a bit. Wilson concedes that while the empiricist worldview will eventually win out against the "transcendental" worldview that people were hoping for, people will still want transcendental beliefs, because *we cannot live without them* (264). At this point Wilson begins to sound like Kidd. From an evolutionary point of view, religion does necessary work for mankind, and so, Wilson says, "if the sacred narrative cannot be in the form of a religious cosmology, it will be taken from the material history of the universe and the human species. This trend is in no way debasing" (265). The Victorian echoes are so loud here (think of Kidd and John Tyndall's materialist mysticism) that it is difficult to imagine Wilson not recognizing them and not recognizing the failure of those prior ideas, which in the wake of the explosion of Darwinian materialism tried to find ways to ease the pain of the loss of spirit in the world.

Given its enormous ambition and its intellectual foundation in evolution by natural selection, *Consilience* is a contemporary utopian spin on Darwin. If Darwin was ambivalent about progress and created a theory that allowed followers a thoroughly nonprogressive reading, Wilson makes the progressivist reading, bridging the difference between Darwin and Spencer. And while Darwin envisioned further studies that would throw light on the human animal, Wilson takes that line and runs to an entirely visionary scientific fulfillment. The enchantment that Darwin's prose offers derives from the complexity, beauty, and sublimity of the natural world; Wilson echoes some of this, but the ultimate enchantment, a self-conscious displacement of religion, is a vision—one can call it no less than visceral and passionate—of universal order, where Darwin's laws are no longer "higgledy piggledy" but the expression of an ultimately unified and coherent world. *Consilience* is a book driven by a rage for order.

The rage, the passion, the faith, and the great intellectual leaps and generalizations make *Consilience*, for me, an astonishingly imperial performance. It echoes with the utopian fantasies of many before Wilson and reverberates with a confidence that is hard to imagine possible in the enormously ambitious, indeed world-historical transformation it proposes. Although I have the deepest respect for Wilson and for the work he has done not only in sociobiology but in the interests of biological diversity, I find his dream more imposing as a dream than as an articulate, rational, and credible sketch for a biologized future. It is a dream that offers itself as a kind of heuristic for research in sociobiology and evolutionary psychology, for the arguments of Dennett and Pinker, and it promises, more than they, an utterly direct connection between biology and culture.

Before concluding this discussion, however, I want to return to the question of whether, given how much political resistance sociobiology has understandably evoked, it entails a commitment to the same sorts of reactionary politics that have marked the history of biologizing the human. Carl Degler points out that in the past, seeing "a biological influence in human behavior" has meant offering "a reason why something cannot be done." So it has been hard, given the history of the uses of biologizing humanity, not to worry about the "harmful ends."

Wilson's strong reductionism, his deep faith that as biological determinants are further explored and taken into account in cultural matters, the possibility of human improvements will be increased, is certainly not intended as a mere naturalizing of the way things are. It is Darwin extended—at least one kind of Darwin. While his own allegiances, in part in reaction to the strong negative response to his earlier work, seem to be conservative—it's hard to know what else Newt Gingrich is doing there in the acknowledgments section (another sign of Wilson's political naïveté?)—that connection is surely contingent.

It is, however, not so widely known that, as Degler points out, sociobiology, like almost all interpretations of Darwinian ideas as they edge over into politics and social action, has been seen as progressive rather than conservative. Degler cites Claude Lévi-Strauss's observation in 1983 that "sociobiology has been taken up by the political left, out of a *'néo-rousseauiste* inspiration in an effort to integrate man in nature,'" while at the same time in the United States "liberals have thrown a 'veritable prohibition over all such research,' a development Lévi-Strauss deplored."[19] In a book with more than Wilsonian confidence in the importance of sociobiology, John Alcock entirely rejects the association of sociobiology with reaction. "At least in my experience," he claims, "sociobiologists are not card-carrying political neanderthals dedicated to defending the status quo and oppressing the masses."[20] Practitioners, he points out, come from the same academic pool that tends "toward the left side of the political equation" (218).

It is likely, in any case, that sociobiology, or some version of it, is here to stay. Insofar as its proponents avoid the extreme reductionism that looks for precise and one-to-one explanations of the fine points of human behavior in genetics and natural selection, it is a project with potential for something other than racism and eugenics. As in the long history of Darwinian influences and uses, the point is not to surrender to the sad history of biologizing but to recognize again the deep dangers of the move from scientific description to moral injunction, and recognize again that however far Darwin pushes us toward a view that biology determines morality, the forces of culture cannot be tamed by biology. "Free play" remains a condition of human life.

It should be clear, looking back over this chapter, that my critique of Wilson is partly a reflex of the similarity between certain parts of his project and my own. That is, Wilson's romanticizing of the natural, the sense his prose often gives (in opposition to the arguments he is trying overall to make) that the natural world is enchanted in spite of its reducibility, is in many ways parallel to what I detect in Darwin's writing. Wilson's passionate secularism is attractive to me, as is the secularism of the great Victorian writers, like Darwin, whom Wilson doesn't seem fully aware he is mimicking. It is the way in which his reductionism translates into an intellectual imperialism that I find most disturbing, and a curious unhistorical sense that pervades even his rendering of scientific history. Darwin's reductionist arguments were often taken, as we have seen, into social Darwinism, into eugenics, and in directions most of us would rather not go.

From here on, I will be giving more direct and sustained attention to Darwin and his writing and will largely be leaving behind the contentious and, often, politically scary arguments I have felt obliged to encounter in these first chapters. But it is critical that readers sustain an awareness of the possible readings of Darwin considered thus far, even as we focus on the remarkably imaginative and—I believe—ethically important aspects of Darwin's writing, and of his life. It would be easy to slip into Wilsonian mistakes and naïveté, or into a complacent sentimentalism about, for example, how pretty the world can look. The "enchantment" I find in Darwin's work is, I hope, something tougher and more aware.

Darwin's reductionism might have led him to the famous line of Dylan Thomas: "The force that through the green fuse drives the flower, drives my green age." This can be felt either as a remarkable and moving imagination of the nature of life, or as a fully reductionist understanding of nature. In Darwin, we find both. "People," he wrote in his notebooks, "often talk of the wonderful event of intellectual Man appearing—the appearance of insects with other senses is more wonderful, its mind more different" (*Notebooks*, 222). He believed it ennobling to recognize one's long historical descent from lower animals and had no fear of this "descent." Reading Darwin one needs to recognize that at least one form of reductionism is compatible with a Romantic view of nature and a deep sense of wonder. Wilson shares some of that Romance, but with a drive to reduce all to the order of scientific law; whatever Darwin's ambitions, however "reductionist" he might have been, there remains the romantic Darwin whom Eiseley noted, a Darwin whose relation to nature fills life once more with the plenitude of enchantment.

CHAPTER 5

Darwin and Pain:
Why Science Made Shakespeare Nauseating

In the first half of this book I have focused more on ways in which Darwin's work has *not* been enchanting than on its power to re-enchant the world. To make the positive argument, it has been essential to face and not to minimize the dispiriting potential of his ideas. In addition, I have attended at some length to those aspects of Darwin's ideas that have been used in support of various theories that might reasonably be called "social Darwinism." It is time to turn to alternative ways of engaging with Darwin, to turn, that is, to the enchanting Darwin, and I want to do so by looking at his writing and at his life, though we know the latter in good measure by the former. I reserve for chapter 7 my most extensive engagement with the qualities of his writing that make for re-enchantment.

Making a biographical case is a tricky business. As I have said at the outset, I concur entirely with those recent historians of science who have denigrated the tendency to write of Darwin in the mode of hagiography. It is not news that Darwin was an unusually nice man for a world-historical figure, but he was hardly perfect. Janet Browne describes the "steely" determination that lay under the gentle and cordial demeanor that marked his relations with just about everyone. She points out, in addition and among other things, that he "could be ruthless in cutting himself off from those to whom he owed the greatest debts" (*Power of Place*, 418). I turn to biographical elements, to a consideration of the ways in which his theory developed and to the way in which he handled a major crisis in his life, not because I like Darwin, though I do, but because these things can serve as useful examples of the way a fully naturalistic vision can be compatible with a sense of the world as value-laden and inspiring. I want to think of Darwin's life not as saintly but as evidence for the

possibilities of "nontheistic enchantment." His way of dealing
with the world might be taken as exemplary of the possibility not
only of doing without transcendent consolation but of finding
positive value in the "merely" natural, and of loving it. Moreover, it is important to my overall argument to have an example
of the way important scientific discoveries might emerge from a
culture-bound consciousness, might in fact be made possible by
just the local and particular contingencies of the life of a scientist
(or artist, or thinker of any kind). Darwin was no saint. He was a
Victorian gentleman. And he managed, being that, to produce
one of the most important ideas in the history of Western culture.

To be fair, I should allow that the strategy of my argument
was anticipated by many nineteenth-century intellectuals (and
by many of their successors), all of whom were happy to grant
that thoroughly naturalistic thinkers might behave with the finest
of consciences and with an almost religious reverence for the
world. Arthur Balfour, in his strong and popular attack on the
authority of science in the areas of ethics and religion, *The Foundations of Belief*, makes an early case for the ethical significance of
"enchantment," though he does not use the word. He lays down
two propositions:

> (1) that, practically, human beings being what they are, no moral code can be
> effective which does not inspire, in those who are asked to obey it, emotions
> of reverence; and (2) that, practically, the capacity of any code to excite this
> or any other elevated emotions cannot be wholly independent of the origin
> from which those who accept that code suppose it to emanate.[1]

Clearly Darwin's imagination of the world lies behind the question of "origins," in Balfour's formulation, and it is precisely to
prove that morality requires other origins than those proposed
by Darwin and by scientists committed to an entirely naturalistic understanding of the world that Balfour wrote *The Foundations of Belief*. It might, he argues, be possible for mankind, "even
instructed mankind," to "preserve injured sentiments which they
have learned in their most impressionable years," in the face of
their Darwinian knowledge that these are "merely samples of
complicated contrivances, many of them disgusting, wrought

into the shaping forces of selection and elimination," but in the long run these "sentiments" will be destroyed, "and the contradiction between ethical sentiment and naturalistic theory will remain intrusive and perplexing" (18–19). The argument is consonant with Weber's: without those feelings of reverence, the world is disenchanted, or will be soon. Thus, for me to argue that Darwin managed to sustain those "injured sentiments" through his own life is *not*, on Balfour's accounting, a real argument for the possibilities of "nontheistic enchantment."

Similarly, W. H. Mallock addresses quite directly the Victorian naturalist's (Mallock calls them all "positivist") argument that naturalists are often good men, of strong conscience. "They have conscience left to them—the supernatural moral judgment, that is, as applied to themselves—which has been analysed but not destroyed." The naturalists claim, says Mallock, that instead of a "foundation of superstition," conscience is set on a foundation of "fact" (145).[2] Like Balfour after him, Mallock concedes that conscience may continue undiminished, "but that alone is nothing at all to the point". And he proceeds with a delightfully effective anecdotal example:

> A housemaid may be deterred from going to meet her lover in the garden, because a howling ghost is believed to haunt the laurels; but she will go to him fast enough when she discovers that the sounds that alarmed her were not a soul in torture, but the cat in love. The case of conscience is exactly analogous to this. (19)

The example, then, of someone who continues to act on the promptings of conscience, or, as I am trying to show in what follows, who continues to live in the world as if it were enchanted, is no evidence that a naturalistic vision might be compatible with an ethical one.

For both Balfour and Mallock the idea that the natural world might inspire "reverence" of the intensity that the "enchanted" worlds of religion inspire is simply unthinkable. In rejecting this possibility, Balfour suggests that the sublime "starry heavens" would be, "on the naturalistic hypothesis," comparable to "the protective blotches on the beetle's back" (18). From the Darwinian perspective, this is absolutely right. And one purpose of this book is just to get readers to recognize the sublime in "the protective blotches on the beetle's back," to recognize and feel it, as Darwin did. I begin here with consideration of Darwin the man (or rather, with some aspects of his life), just because it is critical to any argument for a secularism of this kind, the kind in which the beetle and the worm and the ant and the parasite gnawing at stomach linings, and the bat and the weed and the bird feces that bear potent seeds over continents—all of this is worthy of reverence and can inspire it. If one looks.

The Darwin whose theories have been exploited for various ideological purposes is usually also the disenchanting Darwin. His "dangerous idea" has been felt to be threatening and dispiriting in part because it is inescapable: the mindless, algorithmic processes of natural selection go on all around us, everywhere, create us and our enemies, all the organic world. But the pervasiveness and complexity of those processes have about them as well a quality of the sublime, a "grandeur" that fills the apparently explicable world of material nature with an almost unspeakable richness and beauty.

The alternative to Darwinian disenchantment, however, is not an escape from his theory and his vision but a fuller and more detailed exploration of the way his perceptions and his language might bespeak a secular enchantment. The first step toward a reengagement with the world that Darwin seems so definitively to have evacuated of spiritual meaning and design is recognition of the degree to which his work *and* his life were infused with a sense of the value of things, with a deep emotional engagement in the material world. For Darwin, the organic world to which he devoted a lifetime of attention was irresistibly attractive and beautiful, even when it was most parasitic and destructive; it provided him with the kind of spiritual sustenance that his wife Emma continued to find in traditional religion. Darwin was enchanted by the world. And his world, like the sublime itself, enchants as it also threatens.

I want then to move away from the battles over natural selection and the ideologies to which it has been attached, in order to look at that condition of enchantment that is as central to Darwin's life and work as the sheer intellection that would seem to have produced the theory. While this will continue to require confronting some of the horrors that Darwin knew were always an aspect of the nature he was constantly discovering, it will also help us to recognize that the losses he felt were indeed *felt*, that Darwin was not remotely anesthetic, but deeply (if cautiously) passionate about the world he laid out for us with sometimes chilling dispassion.

In Darwin, intellect and feeling were one. Rather than splitting the world into feeling and reason, a condition of modern disenchantment, Darwin's writing, like his life, reflects a way of thinking and feeling that is distant from the disenchanting austerity of modern interpretations of his work. Early in his *Beagle* voyage, out on a "long naturalizing walk" in Brazil, he seems to have had a revelation: "It is a new & pleasant thing for me to be conscious that naturalizing is doing my duty, & that if I neglected that duty I should at the same time neglect what has for some years given me so much pleasure."[3]

Reading Darwin changes the way one looks at the world, even for us moderns, who have lived all our lives with a post-Darwinian perspective, with the idea of evolution almost a commonplace (despite the fact that it is still denied by fundamentalists and those who offer "intelligent design" as an alternative). Evolution is an idea, but if one spends much time at all observing nature in the wild (or semiwild, or even, for that matter, in the city) one has to become increasingly aware of how complex, subtle, interdependent, and often beautiful adaptations to environment can be, how the very slightest markings can imply important genetic distinctions, and how variations range from clearly marked genera down to the minutest individual difference. It is a shock to new bird watchers, for example, when they discover that a bird they have identified as a differentiated species is suddenly transformed, in the next bird guide, to a mere variant—as happened to me early in my own birding career, when the Slate-colored Junco and the Oregon Junco, lovely sparrow-related and wintry birds, disappeared under the newly defined species Dark-eyed Junco. These seem little things, but speciation is the big Darwinian question, and the intricacies of nature were his subject. The Slate-colored Junco, which shows no touch of color beyond a rich grey and white, breeds, at the edges of its territory, with the Oregon Junco, which shows touches of tan and brown in its feathering. Attending to such minute detail, as Darwin's prose constantly invites us to do, reveals an almost miraculously diversified and changing world. Such subtleties of difference, we know, insofar as they were recognized, were often transformed by natural-theological arguments into evidences of "design," but Darwinian explanations show that the design need not have been "intentional" at all, while the sublime subtleties and complexities do not diminish at all. The miracle is of another kind: that such amazing colors, forms, structures, emerge from natural processes.

There is a striking, characteristic, and wonderfully naïve line in Darwin's *Autobiography*: "From reading White's Selborne," he wrote, "I took much pleasure in watching the habits of birds, and even made notes on the subject. In my simplicity I remember wondering why every gentleman did not become an ornithologist" (45). Darwin watched "the habits of birds," as he would go on to watch the habits of barnacles and worms and climbing plants, and might well have wondered why others were not so completely absorbed by, fascinated by, enthralled by those creatures as well. The ambition to become an ornithologist, to translate the pleasures of bird watching into the work of science, registers simply the complete continuity between Darwin's feelings for nature and his determination to find out everything he could about it. The conventional distinction between labor, conceived by political economists of Darwin's time as always entailing pain, and pleasurable *a*vocation disappears in Darwin's work and in his vision of the world.

The naïve enthusiasm of Darwin's comment, written in his last years, is related to another, sadder quotation from the *Auto-*

biography. Near the end of that book and his life, Darwin famously lamented his loss of feeling for art or poetry: "Now for many years I cannot endure to read a line of poetry: I have tried lately to read Shakespeare, and found it so intolerably dull that it nauseated me" (138). It is as though Darwin himself had felt the pressures of disenchantment that his theory was exerting. But in fact, Darwin's expression of regret about this anesthesia implies a deep feeling about the poetry that now repels him (a strong memory of its pleasures); readers have usually emphasized the loss of feeling here, but emphasis needs also to be put on the fact that the loss is registered feelingly, by someone for whom poetry had indeed mattered greatly, and someone who reacts with instinctive feeling to failures of feeling, not with mere intellectual disinterest, but with an intensity of revulsion positively "nauseating."

The problem is not simply personal: it opens again the question of a more general and culturewide disenchantment. It is this question that has led me to turn to Darwin's life, his own personal response to his own discoveries. Does commitment to a thoroughly naturalistic understanding of the world, does a sense of the world as without divine supervision or any but human spirit, imply a world bereft of value, stripped of its power to enchant? If ethical engagement is ultimately a condition of enchantment, as both Mallock and Balfour imply in their arguments for the importance of reverence, the possibility of secular enchantment becomes essential for secularists like Bennett and Connolly, and like me.

Many critics, especially literary, have speculated over Darwin's apparent aesthetic anesthesia at the end of his life.[4] His revulsion from high art would seem to be related to many other aspects of his life and work. But, as I argue throughout this book, the usual popular understanding of Darwin as projecting a world mechanized, insensitive, reducible to a mere formula of heartless competition, needs to be qualified. As Darwin can open the world for people interested in looking closely at nature, even in the amateur mode of birders, so his way of thinking and feeling about the materials he worked on can entail quite other, less "inhumane" and mechanistic understandings of nature and of what it means to be human. So I want here to suggest, by way of a biographical look at two different phases of Darwin's life, what might account for this anesthesia, and to suggest as well that his rendering of the world, though it is tough-minded and difficult, is derived from and points back toward a humane and loving relation to nature and to people. The worst moment of his life—the death of his beloved daughter Annie—was undoubtedly an important source of his poetic desolation. But that response is most fully understandable in relation to a far different and more famous episode, his enthusiastic voyage into the natural world on the *Beagle*, with the poets always by his side.

I

One of the largest cultural issues concerning Darwin obviously has to do with religion and with what might be taken as the even broader problem of how he and his culture found ways to compensate for the withdrawal of meaning from the world that his overall arguments would seem to have entailed. A world governed by natural selection, the chance collocation of environment and inheritance, itself only chancily developed, would seem a world from which meaning and consolation were withdrawn. A stochastic system, however regular and even formulaic the developments after the unpredictable chance variations, offers only minimal solace.[5] With so little spiritual support left, one would have thought that Shakespeare would have been a big help.

If one thinks of the birds as the nature Darwin spent his life investigating, loving, and desanctifying, and Shakespeare as the literature that late in life he abandoned, one can recognize yet another manifestation of the continuing Western tension between science and poetry. It would seem from such a division that Darwin did participate in the late-century tough-minded

movement of stern and heroic disenchantment, the disenchantment that in the interest of a rationally derived truth required all
thinking people to take a good look at the worst—at a world that
offered no compensations and that remained utterly indifferent
to human aspiration. While it is reasonable to argue, as Robert
Richards has done, that Darwin was ultimately committed to a
view of the natural world as progressive, that certainly has not
been the dominant understanding of the *Origin;*[6] nor could he
ever believe that the general course of things moving toward the
better could ease the pain of deep personal loss. In the most
painful personal circumstances he took a good look at the worst,
but as I have already suggested, he found pleasure in the very
work that would lead him to imagine a world governed entirely
by laws of material nature.

His pleasure in the work of a naturalist was instinctive. The
world was simply fascinating to him. And he could sustain that
fascination and the joy of it without accepting any of the forms
that religion might offer. Darwin is not being disingenuous (as
he might well have been in his conclusion to "Struggle for Existence") when, at the end of the *Descent,* he makes his argument
for the religious implications of his theory:

> I am aware that the conclusions arrived at in this work will be denounced by
> some as highly irreligious; but he who denounces them is bound to shew
> why it is more irreligious to explain the origin of man as a distinct species by
> descent from some lower form, through the laws of variation and natural se-
> lection, than to explain the birth of the individual through the laws of ordi-
> nary reproduction. The birth both of the species and of the individual are
> equally parts of that grand sequence of events, which our minds refuse to
> accept as the result of blind chance. The understanding revolts at such a con-
> clusion, whether or not we are able to believe that every slight variation of
> structure,—the union of each pair in marriage,—the dissemination of each
> seed,—and other such events, have all been ordained for some special pur-
> pose. (2:396)

It is the "grand sequence" that Darwin cannot believe is the result of "chance," though he makes no case for intention or design
at all. But it is the idea that all the smallest details of life (and
death) are somehow "ordained for some special purpose" that he

finds absurd, and that, as would be the case with the death of his daughter, is positively immoral, given the way the world shakes itself out moment by moment, day by day.[7] How terrible to think that the death of his daughter—a sweet little girl of ten years old—was intentional, the work of an omniscient God.

The tough-minded Darwin we have inherited was a softy, perhaps too much of one, both at home and at work. And the dominant views of his work seem to me wrong insofar as they interpret the life he described as nasty, brutish, and short, and insofar as they help in the construction of a sharp division between spiritually uplifting faith and spiritually empty godlessness. For Darwin the science/literature dichotomy does not work; the two are not really distinct and opposed, as the title of this chapter might have too provocatively suggested. The nausea Darwin felt in reading Shakespeare embarrassed him because he was convinced that the failure was his, not Shakespeare's.

The initiating years of Darwin's scientific epos were informed by poetry. Poetry clearly helped shape his imagination and his scientific vision. Gillian Beer in *Darwin's Plots* and more recently Robert Richards, in *The Romantic Conception of Life*, have written extensively about how *Paradise Lost* might have inflected his theories. Poetry was his companion on the *Beagle*, and his first great success as a writer was only partly "scientific." *The Voyage of the Beagle* won attention as a travel book, humanly fascinating and marvelously written, and saturated in Romantic sensibility. Literature and art were for Darwin indications of the remarkable development of human civilization, for willy-nilly he did participate in his contemporaries' tendency to see Western culture as the apex of natural and human development.

Reading Darwin's letters and spending time with the many biographies and biographical narratives about him, I have found it hard not to feel a sentimental fondness for the man, something I have not found easy for many great historical figures and artists. To be sure, he was, almost unself-consciously, a member of the elite and profited in his work and life from that inheritance. Class mattered to Darwin as it did for just about every

Victorian. He was certainly not a man without flaws and prejudices, but all the evidence suggests that he was, for the most part, pretty much a nice guy. The praise of his charm and modesty has in recent years been criticized occasionally as too easily earned, and Darwin's charm is now often seen as a disingenuous effort at self-protection and self-promotion. His illness, similarly, has been understood as a characteristic gesture of Victorian self-defense, and one has to notice that it set in at about the time of his wife Emma's first pregnancy and lingered through much of the work on evolutionary theory that he thought of as like "confessing a murder."[8] But it is easy to be too cynical. Darwin was certainly both modest and compassionate. If he partook unreflectingly in Victorian assumptions about who does the domestic work in the family, he was nevertheless a loving father. And if his position gave him advantages that, for example, Alfred Russel Wallace never had, and a network of alliances that not only assisted him with information but protected him from attack, he never alienated anyone who helped him, was sensitive to those around him, and was clearly both a very good friend and a very affectionate (if patriarchal) husband.

It should be noted that "Unanimously," as Janet Browne writes, "the children rejected their father's own view of himself as a deadened, anaesthetic man" (*Power of Place*, 429). But while Darwin was curiously "incapable of seeing himself as others saw him" (431), his description of his "anesthesia" is an important indicator of his character. The gentle lament for his loss of aesthetic sensibility is a consequence of his long-standing passion for literature and poetry. It marks him as a man who had read literature very well when he was young and who responded to what he saw and discovered in the world with an emotional and moral intensity trained by literature and rare for anyone, certainly for scientists as our contemporary culture often caricatures them.

Darwin's famous voyage on the *Beagle* began with a kind of Romantic exuberance and innocence. Into the ship's cramped quarters, where he was to be captain's companion and only secondarily, after some awkwardness with the incumbent ship's naturalist, he managed to bring some books. Famously, one of those books was the first volume of Charles Lyell's *Principles of Geology*, which had only recently been published. Lyell was an education to Darwin, who looked from the *Beagle* with a uniformitarian geologist's eye, but the dominant presiding figure for Darwin was Alexander von Humboldt, whose *Personal Narrative of Travels to the Equinoctial Regions of the New Continent, during the Years 1799–1804* Darwin received as a gift just before he boarded the *Beagle*. Humboldt had mattered to Darwin for a long time, and that book in particular became the model for much of his note-taking during the voyage and then for his own extremely popular *Voyage of the Beagle*. As Robert Richards puts it, "Darwin's own prose, which vibrated with the poetic appreciation of nature's inner core, had a comparable end [to that of Humboldt's]: to deliver to the reader an aesthetic assessment that lay beyond the scientifically articulable" (521).

But along with the work of these distinguished, cultivated, and aesthetically sensitive scientists, Darwin also took some poetry. He indicated early in his journal of the *Beagle* voyage that along with his scientific pursuits, he intended to study "a little classics," largely the Greek Testament (*Diary*, 13), but with his mind and soul and body devoted to the tropics, Darwin also regularly read Milton. So he wrote, "in my excursions during the voyage of the 'Beagle,' when I could take only a single small volume, I always chose Milton." Gillian Beer argues that "the sustenance he drew" from Milton, in particular, significantly affected "the formation of his ideas" ("Darwin's Reading," 550). It is initially surprising that this great scientist famous for having gone "anesthetic" at the end of his life, and famous for a vision of the world that would seem profoundly anti-Miltonic and brutally unpoetic, would have taken with him everywhere that volume of poetry.

Milton, surely, was part of that enthusiasm that sent Darwin into tropical raptures, that fed his imagination, satisfied his need for beauty and for order, and at the same time encouraged his sense, as Gillian Beer argues, of multiplicity, profusion, and abundance (554). In the wildernesses of South America, maneuvering among genocidal attacks; on the Argentine pampas, where he was forced to eat nothing but red meat for long periods; on the Cape of Good Hope; in Tasmania and New Zealand—everywhere he was finding Milton's imagery compatible with wildness and suggestive of ways to think and feel about it. The one direct reference to Milton in his diary comes after a powerful description of the phosphorescent sea off the coast of Argentina, during a trip toward the Rio Plata: "It was impossible to behold this plain of matter, as it were melted & consuming by heat, without being reminded of Milton's description of the regions of Chaos & Anarchy" (107). Clearly, Milton occupied his imagination of the world around him.

Considering where Darwin eventually came out as a naturalist, it is interesting that he was fascinated by the most famous version of the Western myth of origins outside of the Bible itself. Richards, citing a passage from *Paradise Lost*, takes the point even further, arguing that Milton's vision of the fall, that "out of death and destruction comes life more abundant, life transformed," is "exactly the resolution that nature, in Darwin's divinized reconstruction, offers" (538). On this account, Darwin's scientific argument becomes a literal creation myth, and carries with it the affective weight of myth. Whether or not *Paradise Lost* worked in Darwin's imagination toward the theory he was ready to formulate three years after his return from the voyage, it is evident that poetry mattered to Darwin. He even claimed to have read Wordsworth's *Excursion* twice, a feat rather more heroic, even some literary people might say, than circling the world for five seasick years on the *Beagle.*

Darwin was a Romantic. Long before Captain Fitzroy, on the recommendation of John Henslow, Darwin's teacher and friend at Cambridge, invited him to join him on the *Beagle*, Darwin had dreamed of the tropics. Humboldt's personal narrative infected him with a deep and not very scientific longing. In 1831, before he knew about the possibility of a *Beagle* voyage, he wrote

to his sister, Caroline: "I never will be easy till I see the peak of
Teneriffe and the great Dragon tree; sandy, dazzling, plains, and
gloomy, silent forests are alternately uppermost in my mind. . . .
I have written myself into a Tropical glow" (*Correspondence*,
1:122).

The glow lasted throughout those five difficult and sublime
years, in which Darwin spent much more time on land, facing
dangers and hardships that seem almost incomprehensible today, than he did at sea, where he was always sick. The glow
lasted even after homesickness settled in with the seasickness to
make Darwin long for the soft, green and pleasant land of his
wealthy home in Shrewsbury, where the huge and successful
patriarch, Dr. Robert Darwin, sent out allowances at Darwin's
request. Six months after starting around the world, gathering
the materials that would eventually lead to his theory, he noted
in his diary, "Few things give so much pleasure as reading
[Humboldt's] Personal Narrative; I know not the reason why a
thought which passed through the mind, when we see it embodied in words, immediately assumes a more substantial & true
air. In the same manner as when we meet in dramatick writings
a character which we have known in life, it never fails to give
pleasure" (63).

"What a splendid pursuit Natural History would be," he
once wrote, "if it was all observing & no writing."[9] Yet we know
Darwin through his words, as he understood that his posterity
would know him, and the pains it gave him to get it right, as opposed to the continuing pleasures he always derived from observation of nature, suggest their importance, his sensitivity to
their working and to the traditions out of which they came. The
relatively spontaneous writing of his letters connects him immediately with the romantics, and much of his writing from and
about the *Beagle* is distinctly Wordsworthian. It echoes with a
sense of the wonder of language, and reflects the vision of
Wordsworth, for whom memory could come to displace the immediate moment and transform it. Like a true Wordsworthian,
Darwin intermingled his thought about nature with thoughts

derived from the literature he loved, and with memory. At one point in his diary, he noted: "at present fit only to read Humboldt: he like another Sun illumines everything I behold" (39). Of his landing on a tropical island he wrote: "The delight one experiences in such times bewilders the mind"; "delight is however a weak term for such transports of pleasure" (39).

Like Wordsworth too, Darwin from the start of his voyage was preoccupied with the problem of remembering. He wrote in the original diary of his voyage:

> Excepting when in the midst of tropical scenery, my greatest share of pleasure is in anticipating a future time when I shall be able to look back on past events: & the consciousness that this prospect is so distant never fails to be painful.—To enjoy the soft & delicious evenings of the Tropic to gaze at the bright band of stars which stretches from Orion to the Southern Cross, & to enjoy such pleasures in quiet solitude, leaves an impression which a few years will not destroy. (38)

Even as he revels in the immediate experience of the tropics, he anticipates, as Wordsworth often did, a more satisfying experience yet, the experience of looking back on it.

Darwin's Wordsworthian and Miltonic enthusiasms were not unscientific, however. James Paradis's analysis is surely correct: an essentially poetic response to the natural gives way to—I would prefer to say "generates"—a scientific one. In Humboldt's narrative, facts are always part of the overall Romantic, exploratory impression the book gives. It is not so much that Darwin supplants aesthetic idealism in his narratives as that he develops the Humboldtian commitment to attending to every detail more fully. The *Journal of Researches into the Geology and Natural History of the Various Countries Visited by H.M.S. Beagle* mixes throughout an attractive and affective mode of writing, which registers Darwin's feelings in response to what he sees and experiences, with that more strictly "scientific" record of the facts.

Darwin's writing throughout his career reflects his efforts in describing his experiences during the *Beagle* voyage to register precisely and objectively the facts about the places, peoples, and organisms he observes while rendering the felt sense of those observations. Even the *Origin*, as I will be arguing in more detail later, often begins its engagement with particular facts by registering how astonishing or beautiful they are. Darwin's sometimes awkward adoption of the passive voice suggests, however, that he is always concerned to explain the phenomena in ways that anyone else might also observe, and to see them systematically, as manifestation of "general laws."

The *Journal of Researches* is a totally different book from the *Diary*, transforming the material there, extending it with materials from notebooks and the collections he had gathered, rethinking it, looking at it from the perspective of new thoughts and researches. Darwin's mind was always, from the very start of the voyage, after larger game than the "vast accumulation of facts" at which, according to his autobiography, his mind was always grinding. He had what Philip Sloan has called "synthetic ambitions,"[10] but the diary, of course, is far more personal, far more indulgent of his feelings, than he allows the public version to be. Prosy and unimpassioned though Darwin has so often been taken to be, the *Journal of Researches* is also rich with a sensibility that can only be thought of as poetic. The precision of his observations is rarely without its affect. Here, for example, is a brief passage, in the midst of his discussion of the behavior of shipboard spiders, in which Darwin tries to account for their capacity to travel long distances:

> A spider, which was about three-tenths of an inch in length, and which in its general appearance resembled a Citigrade (therefore quite different from the gossamer), while standing on the summit of a post, darted forth four or five threads from its spinners. These glittering in the sunshine, might be compared to rays of light; they were not, however, straight, but in undulations like a film of silk blown by the wind. They were more than a yard in length, and diverged in an ascending direction from the orifices. The spider then suddenly let go its hold, and was quickly borne out of sight. The day was hot and apparently quite calm; yet under such circumstances the atmosphere can never be so tranquil, as not to affect a vane so delicate as the thread of a spider's web. If during a warm day we look either at the shadow of any object cast on a bank, or over a level plain at a distant landmark, the

effect of an ascending current of heated air will almost always be evident. And this probably would be sufficient to carry with it so light an object as the little spider on its thread. (*Journal,* chapter 8)

It is interesting, reading so splendidly attentive and even literary a passage as this, to think of Balfour's horror at the equating of a beetle's back with the starry skies. Certainly, in such a passage as this, it is impossible not to be awed by the workings of so minuscule and delicate a world, and by the precision of observation in Darwin's rendering. It's a passage that has, then, the kind of scientific rigor Darwin required of himself, but it is also rich with a sensibility that leans toward analogy and metaphor to do important work, and while Darwin here restrains himself from registering admiration and awe, the passion for seeing and the deep attachment to nature are self-evident.

It is not so much, then, that Darwin moved in the *Beagle* narratives from "the aesthetic idealism of Romantic art" to "the system building traditions of geological and natural sciences."[11] Darwin did return from the voyage as "a highly skilled and creative investigator," but he did not, I think, forego "the aesthetic idealism of Romantic art." The world he represents in the narrative had the same kind of appeal to his early Victorian audiences that Humboldt had for him. The thrill of the experience of the tropics and the excitement of discovery, the sense of the wonder of the natural world, remained in Darwin's very acts of observation as he developed those skills in research and analogical thinking that allowed him to produce his theory with some confidence. Throughout his life after the *Beagle,* Darwin focused with the most meticulous intensity on the particulars of the natural world. His objective, of course, was to transcend the personal, as it was manifest regularly in his diary.

But Darwin's experience of wonder is consistent with the sense of wonder that underlies natural theology, and that experience leads him to attempt to discover and then to demonstrate that the wonderful is the product of wormlike activity.[12] Darwin wants to short-circuit the inference from wonder to the transcendent, but not to deny the wonder or its importance. His

reference to Carlyle in his letters suggests something of his dou-
ble relationship to natural phenomena. At one point, in January
1839, he wrote to Emma: "To my mind Carlyle is the most worth
listening to, of any man I know" (*Correspondence*, 2:155) But nine
months later he has become "quite nauseated with [Carlyle's]
mysticism, his intentional obscurity and affectation," to which
he contrasts his own "common Englishman's" mind (2:236). Carlyle invokes wonder to encourage a sense that the world is not
ultimately intelligible to human rationality, but the common Englishman is just a good, solid empiricist, who gets impatient
with obfuscations of the vivid particulars of the natural world.
The wonder for Darwin remains *in* that natural world and in his
experience of it. He insists on seeing the world as intelligible on
its own terms, and as he does so, he still finds it rich with wonder. Darwin splits off from irrationalist Romanticism but remains
intrinsically Romantic still.

We know him best for his great synthetic works, *The Origin of
Species* and *The Descent of Man*, themselves packed with details—
though not enough to keep him from apologizing for not giving
more (he describes the *Origin* as a mere "abstract"). But most of
his work is devoted to detailed studies of minutiae that would
go to support the synthetic arguments: studies of orchids (1862),
climbing plants (1865), variation in domesticated plants and animals (1868), the expression of the emotions (1872), insectivorous
plants (1875), different forms of flowers on plants in the same
species (1877), the power of movement in plants (1880), and
worms (1881). All of these books are informed by the larger theory and designed to provide evidence for it.

Part of the great miracle of his imaginative life was that he
paired in his own temperament an instinctive need to see the
larger picture, to recognize that the details were part of larger
systems, just as Lyell had partly taught him to do, with an instinct for the particular, indeed a passion for the particular, out of
which wonder continued to emerge for him down to his death.
And while he denigrated his powers of feeling in his *Autobiography*, he sustains his Romantic passion for the things of this world

throughout even his "driest" and most meticulously particular-
izing works. He would rigorously refuse (or try to refuse, until
Wallace came along) articulating that larger picture until he was
convinced that none of the particulars he had watched with such
piercing analytic energy would undercut his idea.

His fullest joy at work comes not in the big argument but in the
contemplation of the smallest of things—like the spiders on the
Beagle. At one point in 1846 he wrote to his former sea captain
Fitzroy, "you would hardly believe it if you had seen me for the
last half month daily hard at work in dissecting a little animal
about the size of a pin's head from the Chonos Arch. & I would
spend another month on it, & daily see some more beautiful
structure" (*Correspondence*, 4:359). Darwin's wonder is increased
by dissection. In his work the commonplace and the wonderful
were the same, and the effect of his detailed and intense observations was distinctly Wordsworthian—the dissected barnacles became objects of wonder; even in their minutest details the common realities of nature became wonderful. As Howard Gruber
put it many years ago, one of Darwin's dominant thought forms
is "the summing of small effects over many iterations to produce
large, often surprising results."[13] Wonder was the beginning and
the end of Darwin's research, from barnacles to worms to people.

So Darwin manages, and probably even in those amazing
early days when he seemed to be scooping randomly fish and
shells and vegetation from the sea as the *Beagle* sailed on, to see
the world in a grain of sand. The details always remain intrinsically interesting, even as Darwin's instincts set them up in relationships that ultimately signify something much larger. That
is a poet's imagination, and a great scientist's as well. Rebecca
Stott has explored in admirable detail Darwin's descent into
naming, cataloguing, understanding, all the barnacles known
anywhere in the world, as he pushed on beyond what he imagined as a job of several months to produce, after eight years of
research, three monographs that became *the* authoritative study
of barnacles[14]—all were only possible because he was working
with a theory that he was afraid to, not ready to, publish.

The poetic mind and the analytic, the theorist and the dis-
sector: Darwin was each and all, usually at the same time. The
Origin, he claimed, was one long argument, although many lay
readers find it too densely packed with detail. The larger conception and the particulars are aspects of each other, and in
Darwin there is the strongest confirmation that the common
assumption—usually inferred, as well, from the "anesthesia"
Darwin reports in his autobiography—that Romantic passion is
incompatible with analytic attention is simply wrong. The *Journal of Researches*, crammed with facts, on Humboldt's model, is
also crammed with Wordsworthian reflections on and responses
to nature and to memory. That book does begin to raise the question: what has poetry to do with the rough and inhuman landscapes of a nature that seemed not to care about consciousness,
memory or desire, at least not about *our* consciousness, memory,
and desire.

However bleak Darwin's world has looked to many people,
Darwin's Romantic enthusiasm for nature in all its often grubby
or nasty detail never diminished, through Shakespeare-nausea
and prolonged illness and international fame down to his last
book, in which he lavished affection and the most patient, infin-
itesimal attention on worms. *The Formation of Vegetable Mould
through the Action of Worms* might be taken as a kind of panegyric
to worms, whose intelligence, skills, and digestive tracts link
them ultimately to humanity and raise and swallow empires.

Darwin does not rhapsodize about worms. One needs to speculate, however, on the sort of imagination that allowed him to
get so down and dirty with them. Tennyson seems to have
drawn on Darwin for an image he uses of worms intelligently
pulling the narrow end of a leaf first into their holes. But Darwin's prose is far from Tennysonian:

> Elongated triangles were cut out of moderately stiff writing paper, which
> was rubbed with raw fat on both sides, so as to prevent their becoming ex-
> cessively limp when exposed at night to rain and dew. . . . As a check on the
> observations presently to be given, similar triangles in a damp state were
> seized by a very narrow pair of pincers at different points and at all inclina-

tions with reference to the margins, and were then drawn into a short tube of the diameter of a worm-burrow.[15]

The details of the experiment proceed, in a prose meticulously impersonal and passive in form. Who were these experimenters? Undoubtedly all of this happened in Darwin's study or his garden with the assistance of some of his by now mature children. Were they having fun? I think so.

In observing the worms at work on his little triangles, Darwin is equally meticulous and impersonal in his prose. The results, however, preclude chance: "of the 303, 62 per cent had been drawn in by the apex. . . . As the case stands, nearly three times as many were drawn in by the apex as by the base. . . . We may therefore conclude that the manner in which the triangles are drawn into the burrows is not a matter of chance" (89). In Darwin's intense analysis of all the material he gathered from these and other experiments, it is clear that he is trying to get into the mind of the worm. He is convinced of the continuity of being between worm and human, although of course this never comes immediately to the surface; it is clear that he tries to attribute to worms, and finds experimental sanction for it, qualities of real consciousness and the capacity for choice. "We can hardly escape from the conclusion," he says, in a formulation common throughout his works, "that worms show some degree of intelligence" (93). "It is surprising that an animal so low in the scale as a worm should have the capacity for acting in this manner" (95). Not exactly poetry, but the whole passage has the revelatory power of a poem, as the large sale of Darwin's last book seems to suggest.

And there is a final irony here, for some time after he wrote in his autobiography that he had lost his feeling for art, he tests out the artistic responses of worms. In the very first chapter, Darwin is concerned with "the habits of worms," and he determines that they "do not possess any sense of hearing" (26). But, he discovers, while they did not respond to the playing of a piano, no matter how loud, when they were not *on* the piano, when they were on it "and the note C in the bass clef was struck," they "instantly retreated into their burrows" (27). They retreated again "when G above the line in the treble clef was struck." So Darwin wants to discover if the worms too are anaesthetic—and they are not.

Darwin's extraordinary curiosity about the talent of worms has to do with his inveterate anthropomorphism; this anthropomorphism was, as I will try to demonstrate more fully in my discussion of the theory of sexual selection, absolutely central to his large theoretical project, which attempted to reveal the complete continuity of life through all organisms. The science here was and remains the poetry, for his poetically inspired imagination, no matter how muted and impersonalized, animates and personifies. Out of the infinite details of animate being, Darwin constantly finds fascinating possibilities that only someone attentive to minutiae and conscious of their broadest implications could discover. Darwin's world is not dead and insensitive, not merely mechanical. Its vitality, abundance, complexity, richness are, rather, almost overwhelming in their possibilities, and the world reverberates with sensibility, even the worms' world.

Darwin doesn't, like Shelley, in his powerful poem on Mont Blanc, indulge the terrors of the sublime. By the end, there seemed to him quite enough terror in the nature that he loved. He didn't need it also in poetry. But his very efforts to *de*sublimate the sublimities of creation create an alternative sublime. Of Porto Praya, he wrote in *The Voyage of the Beagle*, the island would "generally be considered as very uninteresting; but to any one accustomed only to an English landscape, the novel aspect of an utterly sterile land possesses a grandeur which more vegetation might spoil."[16] Seeing the world self-consciously from the perspective of what he knew and what he read seduced him into longing for what was not like home: "sterile" landscapes were suddenly sublime for him.

His evolutionary theory notoriously decentered the human. Evolution doesn't happen "for" human beings but works impersonally and inhumanly, unintentionally, without "design." This is familiar to everyone. But that decentering—another aspect of his modesty—allowed him to hear the voices of what was not him or his culture, and he refused to make proud claims for himself or for his species. This was an aspect of virtually every part of his life and work. And if it seems austere, it needs to be understood within this larger vision by which the whole world is animated, and his larger sense that tough as the process is, natural selection evokes the good out of the bad, moves us on a path parallel to the Christian one that he was ultimately to abandon—through a fortunate fall *in* nature, not under God's direction.

One of his first public challenges to poetry comes early in *The Voyage of the Beagle*, at a time when poetry was surely still very important to him. His commitment even then to meticulous observation of the facts led him to notice, even with enthusiasm, the minute, the sordid, the unpleasant. He begins by noticing the "brilliantly white colour" of St. Paul's Rocks, which, he explains, comes from the "dung of a vast multitude of sea fowl" (47–48). As Darwin was later to account for human moral and aesthetic senses by deriving them from the sexual and pack instincts of social animals, so here he transforms the aesthetic into something that, in the first instance, at least, would seem equally disenchanting.

How, Darwin was already asking himself, did life begin—here, in this place? How did vegetation and fauna ever arrive on the relatively new volcanic island of St. Paul's? He lists the few terrestrial fauna that live there, noting

> A fly (Olfersia) living on the booby, and a tick which must have come here as a parasite on the birds; a small brown moth, belonging to a genus that feeds on feathers: a beetle (Quedius) and a woodlouse from beneath the dung; and lastly numerous spiders, which I suppose prey on these small attendants and scavengers of the waterfowl.

No suggestion here that any force other than a natural one might have put life on this blank slate of an island. And Darwin concludes by rewriting the narrative of his beloved tropics in a way that would seem implicitly to undercut his enthusiasm for them.

> The often repeated description of the stately palm and other noble tropical plants, then birds, and lastly man, taking possession of the coral islands as soon as formed, in the Pacific, is probably not quite correct; I fear it destroys

the poetry of this story, that feather and dirt-feeding and parasitic insects
and spiders should be the first inhabitants of a newly formed oceanic land.
(*Voyage,* chapter 1)

But the antipoetry has, of course, its own poetry, and, as opposed
to the falseness of the narratives of conventional "poetry," Darwin offers a narrative that does not in fact diminish his enthusiasm. It gently rejects conventional poetic descriptions ("not *quite*
correct") to replace them with what would be least humanly satisfying, if one were anthropocentric (as opposed to anthropomorphic). Darwin is always tough on theories, like natural theology,
that stand on points of pride and hierarchy. At no moment in
his career—and this passage makes that clear—was he much
alarmed at finding that the human was allied to other, less "civilized" organisms, or that the traditional hierarchy might be challenged by what were often called "lower forms." The anthropocentric poetry that imposes a human charm on tropical islands
is merely false. But there is excitement in a barren place like
St. Paul's, precisely in that it reverses the conventional narrative
to reveal what many would take as the repellent truth: that even
Darwin's beloved tropical forests owe their life to lice, parasites,
spiders, flies, moths, and dung. The previous "scientific" explanations are reduced to "poetry" because they have been distorted
by the requirements of human aesthetic pleasure. Darwin's reverse myth of origins is, however, almost more staggeringly sublime, because while "in the beginning was the parasitical bug" is
hardly as ideal as the biblical "beginning" with God, or, in The
Gospel of John, with the "Word," the emergence from those bugs
of a whole world of tropical wonder is truly miraculous.

Darwin's entire life's work is in a way a long story of exactly
this kind, teaching us to decenter the human but at the same
time not to feel in that decentering a loss or a degradation. It is
no shame to have been preceded by a woodlouse. And the world,
seen from that perspective, is suddenly animated with human
values, even in the activities of parasitic spiders.

Take, as an example of the decentering strategy that Darwin
quickly developed, his discussion of the eye in the *Origin.* Darwin

builds a narrative that resists the anthropocentrism of natural theology by explaining the mechanism of the eye as the product of a mindless and random series of developments, worked out under the control of natural selection. He reverses William Paley's analogy with the telescope. Paley had argued that since there can be no question that the telescope was created intentionally by a designing mind, so a fortiori the eye was developed by a designing mind, that is, by God. Darwin, however, reverses Paley's argument by insisting that to reduce biological possibility to so crude an intentionalist mechanism is impious. It is the clergyman who is irreverent. "May we not believe," Darwin asks, "that a living optical instrument might thus be formed as superior to one of glass, as the works of the Creator are to those of man?" Putting aside Darwin's apparent disingenuousness here in invoking the Creator in this implicit rejection of him, notice how he argues that it is the *imagination* that limits the human sense of the possible, not reason.

It is imagination, so persistently human-centered even in its most extravagant creations, that gets in the way of believing in the possibility that natural selection could produce so remarkable a "contrivance." The skeptic about natural selection is not unimaginative; he is unreasonable. "His reason," says Darwin, "ought to conquer his imagination" (188). Reason, it turns out, is far more imaginative, in our sense of the word, than imagination. Poetic imagination gets in the way of good science, which can conceive of worlds far stranger than poetry's. Darwin was detecting imagination's poverty before nature, and before the reason that was now revealing more of its secrets, even its irrationalities. It might follow from this, in Darwin's way of thinking, that the fullest rational understanding is the understanding most filled with awe.

Nevertheless, as a scientist his perspective was pretty clearly descended from poetry, as he struggled to see everything anew. Under Darwin's eyes, and with the same sort of Romantic impetus that led Blake to see the world in a grain of sand, the ordinary, like the worms, becomes strangely unfamiliar: the stable world seems to spin into motion, and all things—rocks, ants, barnacles, corals, climbing plants, pigeons—reveal their unique histories. Learning to see what had hitherto been invisible out in the tropical wilderness, Darwin was finding all of nature—even, eventually, those home-grown worms—exotic and sublime. He was learning on the *Beagle* to move through and beyond the poetry he knew and loved, in preparation for what he didn't then know would be his ultimate scientific transformation of the world.

One of his great virtues as a naturalist on the *Beagle* was that everything he observed, and he seems to have observed everything, offered itself to him as a question. He had learned much of this approach from Humboldt, whose model he followed almost to the smallest detail. But it is difficult, even with such training, to know what questions to ask, to turn a random fact into an interesting phenomenon that might point toward a higher generalization. Where did this come from? How did it get here? What is it composed of? What might it be compared to? The questions are never cosmic, never transcendental, and the answers are always material. The result, miraculously, is the development of so rich and various and beautiful a world as the one he was sailing through on the way—we may Whiggishly imply—to his grand theory. The trick was to note everything, gather the material to reflect on.

This startling new vision of a world that might, as he noted, be another planet was only partly domesticated by Darwin's experience of literature. Wordsworth had found consciousness in nature, the mind and memory turning the material into humanly significant life. Although Darwin may have seemed to have banished human consciousness from the workings of the world and had no illusions about how nasty inanimate and even animate nature might be, the whole thrust of his work was to discover, as he did in worms, the presence of consciousness; to take a very nonscientific view of his project, it feels like a massive Wordsworthian effort of humanization, of personification. The worms have built kingdoms intelligently, even if under the pressure of the most primitive biological urgencies. Bees devise the optimum solution to storage problems. Nowhere in the animate world, down to its smallest, least intelligent denizen, is it barren of consciousness, desire, and intention. And nowhere, even in that wasteland of Porto Praya, does human consciousness fail to find the associations that charge the world with value and significance.

Like Wordsworth, Darwin during those five years circling the world, reflected on the possibility of reflection. Even as he recognized the impossibility of tranquility amidst the wonders he was experiencing, the absence of associations that ultimately fill the world with significance, he thought of the very different joys of a world full of the familiar. As he registers the intensity of the new, he moves to the edge of a sublime beyond Wordsworth's world of associations; and at the same time he reflects about the alternative kind of beauty and about the possibility of reflecting on this new experience in a different way once it was in the past:

> Many of the views were exceedingly beautiful; yet in tropical scenery, the entire newness, & therefore absence of all associations, which in my own case (& I believe in others) are unconsciously much more frequent than I ever thought, requires the mind to be wrought to a high pitch, & then assuredly no delight can be greater; otherwise your reason tells you it is beautiful but the feelings do not correspond.—I often ask myself why can I not calmly enjoy this; I might answer myself by also asking, what is there that can bring the delightful ideas of rural quiet & retirement, what that can call back the recollection of childhood & times past, where all that was unpleasant is forgotten; until ideas, in their effects similar to them, are raised, in vain may we look amidst the glories of this almost new world for quiet contemplation.

John Ruskin, with a yet more intense Wordsworthian sensibility, felt that a world like this, without human associations, would throw a pall of blankness and chill over experience. As he wrote in some "Academy Notes" of 1857, "Into all good subjects for painters' work, either human feeling must enter by some evidence of cultivation, or presence of dwelling-place, or of ruin; or else there must be some sublime features indicative of the distress as well as the beauty of nature."[17] And yet more intensely, he argues in *Modern Painters*: "Where humanity is not, and was not, the best natural beauty is more than vain. It is even terrible; not as the dress cast aside from the body; but as an embroidered shroud hiding a skeleton" (7:258). Darwin, recognizing the beauty but not being able to respond to it, is being romantically Ruskinian, and Ruskin of course was being romantically Wordsworthian. (It is striking that later Ruskin viscerally recoiled from Darwin's theories, and yet they were friends of a sort and would walk together into the hills of the Lake Country, sharing at least an admiration for the landscape even as they diverged radically on other matters.)

In encountering a restlessness in himself that he cannot quite understand Darwin lives out the Wordsworthian imagination of things, but he works through it differently. He understands the concept of recollection in tranquility, and he feels the sublime difficulty of a world without deep human associations. And yet here is Darwin finding in such a world a delight, precisely without calm, without earlier ideas, but only in contrast to Wordsworthian retrospect. So Darwin's new world is not emptied of the joys of poetry; it becomes comprehensible only in relation to the poetic experiences he (and Ruskin) had, surely, partly learned from Wordsworth. A kind of Miltonic sublime pushes him riskily beyond emotion recollected in tranquility, even as that emotion remains attractive and self-consciously contrasted with the tropics he could not, it seems, get enough of. If he found himself incapable here of experiencing tropical nature as Wordsworth's poetry had taught him nature should be experienced, his mind, "wrought to a high pitch," gave him a pleasure beyond poetry that can only be thought of as the experience of the sublime: not emotion recollected in tranquility, not contemplated in retrospect, but pressingly, inexplicably there, full of anxiety, danger, and incomparable joy.

But Darwin was never to be entirely happy with the mindlessness and human indifference of the world he delighted to observe, and Robert Richards has argued that his deep Romantic roots in *Naturphilosophie* gave Darwin quite another nature than the one predominantly received in post-Darwinian culture. "The nature that Darwin experienced," Richards claims, "is not a machine, a contrivance of fixed parts grinding out its products with dispassionate consequence" (525). It was, simply, alive. He would never again experience the risky and overwhelming sublime of his voyage years, and he would encounter in his studies the brutal strategies of nature that would seem to entail either a rational rejection of it as a model for the human or an elaborate excuse for human wrongdoing. But the world remained fascinatingly alive, not a mechanism but an organism, in the dominant romantic tradition. Alive and difficult, beautiful and dangerous, it was nevertheless, his extended narrative comes to insist, the source of morality and of art.

His consolation for its apparent indifference to human values and feelings, that sometimes the war of nature slows down, that many organisms die without pain or fear of death, and "that the vigorous, the healthy, and the happy survive and multiply," seems feeble indeed—and surely would seem moreso if one were among those not so vigorous, like some of his own children, offspring of a marriage of cousins.[18]

Darwin wrote extensively about, as one chapter in *The Variation of Animals and Plants* calls it, "The Evil Effects of Close Interbreeding." Pretty certainly written as early as 1857, one passage talks quite explicitly about how "the consequences of closer interbreeding carried on for too long a time, are, as is generally believed, loss of size, constitutional vigour, and fertility, sometimes accompanied by a tendency to malformation" (2:90). In his own life, faced with the hard reality of nature's hostility to crossing (like the Darwin-Wedgwoods), Darwin's belief that natural selection allowed "the vigorous, the healthy, and the happy" to "survive and multiply" did nothing to console him for deep personal loss, for lives of pain, for death. Yet however feeble his apology looks, it is serious as well. There are no supernatural consolations.

Clearly, the joy of natural history, the deep pleasures that his studies gave him, were revealing some things that were entirely too painful. Yet, perhaps ironically, it was the pleasure of the work, and the capacity to find a nature so relentlessly indifferent to individual desire, that sustained him through the deepest of difficulties. Although from the start, on his *Beagle* voyage, he rejected the easy "poetic" vision of the world for the demystified and yet sublime encounters with it that close attention required of him, he lamented his loss of feeling for poetry; and he did not harden to a world in which "the war of nature" seemed to be sanctioned by the natural processes he was trying to understand and from which his beautiful and beloved daughter could be pointlessly removed. John Bowlby discusses at length Darwin's difficulty in bearing other people's suffering, his "horror of cruelty and fear of his own anger." These qualities, Bowlby goes on, which "endeared him to relatives, friends and colleagues," were "developed prematurely and to excessive degree."[19]

II

Certainly since James Moore wrote about the death of Darwin's daughter, a great deal of attention has been shifted to that most terrible event in Darwin's life.[20] I had stumbled on the story early, reading through the fifth volume of the enormous and excellent Cambridge edition of Darwin's letters (now through thirteen volumes and counting), and I found myself quite literally crying and having to read some of the passages to my wife. While Moore is not at all sentimental about the awful event, he was certainly right that it was pivotal for Darwin and, I believe, for our understanding of him in relation to large cultural issues. Like many of the passages in the diary of his voyage on the *Beagle*, the story of his relation to his daughter gives us quite another Darwin from the mechanizing, disenchanting figure the culture largely knows.

Emma Darwin was so advanced in her pregnancy that it was impossible for her to travel to the sanitarium to which she and Charles decided to carry the ailing Annie (and where Darwin himself used to take his water treatments). When Darwin followed Annie to Malvern, he committed himself to making his letters to Emma do what, in our times, we would do with regular phone calls, or now, perhaps better, emails: Keep her informed up to the moment, with an urgency that entailed several messages a day. The keenly observant scientist, always alert to minutiae, brings his trained eye to bear on his beloved daughter. The letters thus recount Annie's progress or decline minutely—her vomiting, her pulse rate, her diet of gruel with brandy every half-hour. This "is much bitterer and harder to bear than I expected—Your note made me cry very much—but I must not give way & can avoid doing so, by not thinking about her. It is now from hour to hour a struggle between life & death" (*Correspondence,* 5:14). True to the habits he had been for several years developing in his attention to barnacles, Darwin relentlessly (and lovingly) registered all the details of his daughter's decline: "four deluges of vomiting she has had today—poor thing"; "you could not in the least recognize her with her poor hard, sharp pinched features. I could only bear to look at her by forgetting our former dear Annie. There is nothing in common between the two" (5:16); "3 oclock—the Dr says she makes no progress . . . 4 oclock We have been trying an injection with no success . . . 5 oclock just the same. I will write before late Post" (5:17).

He tells Emma that he thinks it "best for you to know how every hour passes. It is a relief to me to tell you: for whilst writing to you, I can cry, tranquilly" (5:18). It is this image, of Darwin writing with the precision to which he was always committed, and finding that the very experience of writing was an emotional relief from unbearable feelings, that should help revise our sense of Darwin, the bringer of disenchanting news to the world. It is not only here, but it is most obviously here, that the "facts" are laden with (normally unspoken) feeling. And so he tells Emma about Annie's vomiting, her feeble attempts to eat; the bath he and his sister-in-law Fanny gave her with "vinegar and water" (5:19); the reducing of the odor of vomit—"we keep her sweet with Chloride of Lime" (5:20); and then, on April 23, the very day of Shakespeare's birth, "She went to her final sleep most tranquilly, most sweetly at 12 oclock today. Our poor dear child has had a very short life but I trust happy, & God only knows what miseries might have been in store for her" (5:24). The Darwins were not thinking about Shakespeare at that moment and instead we find again the not quite convincing consolatory voice of "Struggle for Existence." It is the only consolation Darwin can think of, and it has to do with this world, not the next. Emma responded, talking of her longing for Annie and her immediate indifference to the other children, and telling Charles, "You must remember that you are my prime treasure (& always have been) my only hope of consolation is to have you safe home to weep together" (5:24).

Emma was right to have worried about him. Darwin wept inconsolably and got sick himself immediately, and he did not attend his daughter's funeral, the details of which were taken care of by his sister-in-law, Fanny, who had been with him throughout those awful last days. Two days later, he writes to his brother, Erasmus, asking him to insert a notice in the papers: "At Malvern on the 23d inst; of Fever, Anne Elisabeth Darwin, aged ten years, eldest daughter of Charles Darwin Esq. of Down Kent" (5:27). By early May, his correspondence with scientists all over the world, primarily related to his work on cirripedes or barnacles, resumed in great detail.[21]

The letters of that awful week reveal many of Darwin's characteristic attitudes. He tells Emma that he "must not give way & can avoid doing so by not thinking about [Annie]." But he seems not to have stopped thinking about her for a moment and to have attended to every fluttering of her pulse. The consolation from that week was for both Darwin and his wife the record of his relentless observation of the fluctuations in his daughter's symptoms. He tells Emma that "it is a relief for me to tell you . . . how every hour passes." Emma replies, "your account of every hour is most precious" (5:21). And she asks him, now aware that her daughter is dead, not to throw away an optimistic letter of that day that he did not send. "I shall like to see it sometime" (5:24).

His daughter's pain had been unbearable to him, and yet he was compulsively and lovingly driven to confront it. Both sensitive father and husband and recording scientist are present in these letters. But for the death of his daughter there was no real consoling explanation. After the careful recording of the slow death, Darwin had to turn away from the grave and return to his loving wife, whom he loved much in return. He was with her when, three weeks later, she gave birth to Horace, their ninth and last child. But nothing could make the death meaningful, and both he and Emma admitted to bitterness. It seems that within two days of his daughter's death, Darwin, with the compulsion of a scientific man used to registering his experiences in diaries and notebooks, wrote a moving memoir of Annie: "I write these few pages, as I think in after years, if we live, the impressions now put down will recall more vividly her chief characteristics" (5:540).

The memoir is both a rigorously careful attempt to register those characteristics, the kind of writing to which Darwin was already much habituated, and a kind of ode to joy, to Annie's perpetual exuberance, "her buoyant joyousness" and her "animal spirits," which "radiated from her whole countenance rendered every movement elastic & full of life & vigour." The deeply personal nature of the document makes it in many ways a bad model for Darwin's more scientific writing, but what is striking about it is the way the scientific and the affective are unselfconsciously blended. Here more intensely, of course, but the memoir suggests again how for Darwin the world described is never simply dead and spiritless, and how his work as scientist is never divorced from his passion for the natural world. If what he is describing is infinitely painful (as is so much of the natural world that appears in his books), it also matters enormously; Darwin cares about it, but the caring is not allowed to influence his recording of its condition.

Darwin carefully notes the development of Annie's "sensitiveness" from a very early age; characteristically, he traces also her "strong affection" from the time she was a baby, when she was

never easy without touching Emma. While there is no more personal and moving document in the Darwin archives, it is touched
throughout with his scientific habits and interests. So he notes
that "Her figure & appearance were clearly influenced by her
character" (5:541). Like the naturalist on the *Beagle* he notes every
fact he can remember: the way she held herself upright, her
height in relation to her age, her "slightly brown" complexion,
her "dark grey" eyes, her large white teeth, her long brown hair,
the difference between the daguerreotype and the living girl. Of
course, there are no negatives here, but Darwin's eye seems
trustworthy. In this document, at least, there is no attempt to disguise the feeling that impels the observation. The last line of the
memoir: "Blessings on you."

And where might those "blessings" come from? Darwin was
not worried then about the possible irony of his implicit invocation of religion. But the hard experience of Annie's death certainly had larger implications for his attitude toward religion, as
James Moore has argued in his essay on the subject. In the *Autobiography*, there is a telling passage that Emma and the rest of the
family decided to excise from the original publication:

> That there is much suffering in the world no one disputes. Some have at-
> tempted to explain this in reference to man by imagining that it serves for his
> moral improvement. But the number of men in the world is as nothing com-
> pared with that of all other sentient beings, and these often suffer greatly
> without any moral improvement. A being so powerful and so full of knowl-
> edge as a God who could create the universe, is to our finite minds omnipo-
> tent and omniscient, and it revolts our understanding to suppose that his
> benevolence is not unbounded, for what advantage can there be in the suffer-
> ings of millions of the low animals throughout almost endless time? This
> very old argument from the existence of suffering against the existence of an
> intelligent first cause seems to me a strong one; whereas, as just remarked,
> the presence of much suffering agrees well with the view that all organic be-
> ings have been developed through variation and natural selection. (90)

Natural selection tells a dirty story. And if it's not natural selection that is doing the dirty work, then it is God, and who, Darwin
wondered, could worship a God like that? Remembering Annie's
pinched, drawn face, he was revolted by the very thought.

Darwin's most extended and theoretically significant moral and intellectual impulse was against the argument from design. The death of his daughter made that argument not only intellectually impossible but morally repulsive. William Paley's *Natural Theology* confronts the imperfection of the world, but provides compensation rather as Darwin did in "Struggle for Existence," that is, impersonally, almost, one might say, statistically. The world, Paley explains, "is not a state of unmixed happiness," nor a state of "designed misery." It is, rather, "a condition calculated for the production, exercise, and improvement of moral qualities, with a view to a future state."[22] Paley recognizes that chance plays a major role in this world, despite the fact that the world is demonstrably a product of design. To deal with the "chance" that develops within the larger design, Paley falls back on a general consolation and an appeal to a larger, transcendent concern. For Darwin, such extranatural, and such generalized, compensation, was no compensation at all and revealed an all-knowing God as utterly cruel. Darwin could not stand other people's pain, and he could not imagine a God who could allow that pain. Perhaps ironically, it is the theologian who offers a rational and disenchanted world; it is the "devil's chaplain," the nonbelieving natural historian who feels the pain of the individual, and who will not allow justification of the death of his daughter, who offers us an enchanted one.

No conception of an all-caring God could justify his daughter's death, or the manner of that death. Darwin's deep Romantic engagement with nature did not simply empty it of spirit and turn it into a mechanism; he pushed toward some sense of amelioration and meaning. But if Annie's death were part of some large, morally significant design, that design was outside the realm of any beneficent and caring deity. Darwin, like Einstein, did not believe that chance could play an important part in the workings of nature, and he self-consciously resisted the disenchanting implications of a world from which the supernatural had been entirely removed. The special strength of his work is precisely in his unblinkered recognition that natural selection, a relentless process that "tends" the individual, entails as well the perennial suffering of animals (and the death of his daughter).

He couldn't reason or imagine his way out of the harsh, natural fact. He confessed at times that he was in a "muddle" on the subject. And yet when he wrote in 1860 to Asa Gray, the American supporter of the *Origin* who nevertheless read it as being compatible with the argument from design, he concluded:

> I see a bird which I want for food, take my gun & kill it, I do this *designedly*. —An innocent & good man stands under tree & is killed by flash of lightning. Do you believe (& I really shd like to hear) that God *designedly* killed this man? Many or most persons do believe this: I can't & don't. —If you believe so, do you believe that when a swallow snaps up a gnat that God designed that that particular swallow shd snap up that particular gnat at that particular instant? I believe that the man & the gnat are in same predicament. —If the death of neither man or gnat are designed, I see no good reason to believe that their *first* birth or production sh^d be necessarily designed. (*Correspondence*, 8:275)

It is difficult here not to be reminded of Gloucester in *King Lear*, saying, "As flies to wanton boys are we to th'gods. / They kill us for their sport."

That Anna died on Shakespeare's birthday is a coincidence (is "intelligent design" an option?) of which I want to take advantage, as I return to Darwin's comment that Shakespeare had come to nauseate him. It wasn't, I'm sure, because Anna died on the 23rd of April. I have been making two potentially opposed arguments about Darwin's resistance to poetry in his later years. On the one hand, I have tried to show that his reflections on his voyage led him to reject poetry as being too satisfying to the human, as in his comments about St. Paul's Rocks. He seems to have resisted in the end Wordsworthian poetry's tendency to make all of nature reverberate with human associations. In this respect, poetry simply falsified and was not consonant with the materialist understanding of nature to which his observations and reflections were driving him. Poetry seemed to share with natural theology an anthropocentric understanding of the natural world. On the other hand, I have argued that Darwin himself was horrified and sickened by sight of suffering, and particularly of suffering that seemed to him meaningless and unjust. Why then would he not turn to poetry as a resource against the brutal indifference of the material processes of nature?

Because poetry of the sort that Darwin didn't want to read any more, Shakespeare's poetry, did not inevitably offer the sorts of humanly satisfying narratives Darwin rejected in natural theology. (Shakespeare, I would argue, could have found poetry in Darwin's discovery that parasitic insects were the beginning of life on barren tropical islands.) Those kinds of narratives would have been, as he put it frequently in the *Origin*, "fatal to my theory." But what effect on his theory, on his thoughts about his daughter, would there have been, say, if Darwin had forced himself to read the passage at the end of *King Lear*, in which Lear holds dead Cordelia in his arms and watches her with Darwinian attentiveness? I simply can't imagine him doing it, and not because he would have found it dull.

> Howl, howl, howl! O, you are men of stones.
> Had I your tongues and eyes, I'd use them so
> That heaven's vault should crack. She's gone for ever!
> I know when one is dead, and when one lives.
> She's dead as earth. Lend me a looking glass.
> If that her breath will mist or stain the stone,
> Why, then she lives.
> This feather stirs; she lives! If it be so
> It is a chance which does redeem all sorrows
> That ever I have felt.

But Annie and Cordelia are dead, and there is no redemption of all sorrows.

The unbearable pain of others was sufficiently widespread through the nature he discovered and described that he didn't want to confront it in poetry, or outside the work that he could never stop doing. Poetry either lied by giving nature a sympathy his investigations could not detect in it, or lied by redeeming and consoling for losses that were meaningless and unconsolable. And when it treated of these losses, it, like Annie's death, was unendurable.

Darwin was too much the poet to endure poetry or Shake-speare. He was too much the scientist to believe that he could make sense of the world by imposing on it his own, or his species', emotional needs. In the *Origin* he personifies natural se-lection as an intelligent being infinitely more perceptive than humanity and careful of the individual to which it "tends." But under the pressure of critics who saw "natural selection" as an active force, actually producing variations, and personified as a living being, he became careful in later editions to remove the Romantic, loving figure, tending to its subjects, and to explain it in the driest language he could find: "I mean by Nature, only the aggregate action and product of many natural laws, and by laws the sequence of events as ascertained by us."[23]

Richards has provided convincing evidence that inspiring Darwin's scientific understanding of the workings of nature there was a strong Romantic impulse that pushed him toward a sense of an ultimate meaningfulness and beneficence. The very metaphors he sometimes tried to replace are evidence of this an-imating spirit in his science. But at this nervous moment, and as his career moved toward an end, his language begins to sound a little like the language of "algorithm," as Dennett describes the process of natural selection. Stripping nature of the support of his marvelous metaphors, Darwin is forced to confront the meaninglessness and injustice of Annie's death. Only by remov-ing "natural selection" from the world of metaphor, which it in-habited in all the early versions of his theory, Darwin in effect gives up on poetry as well, and only then could he live with it. There may be ultimate consolations in the large movements of the world, but the real consolation, the redemption of sorrow, the resurrection of Annie—that was not possible for him, and the pain was too acute for false cheer.

So, of course, it wasn't science itself that made Shakespeare nauseating to Darwin. He never could have imagined that science and literature would be incompatible, and his career reveals a mutual shaping of those forces. Science made literature unen-durable because science was already deeply informed by the moral implications and emotional intensities of literature.

Yet as James Moore and Adrian Desmond emphasize, Darwin is buried in Westminster Abbey, and "Getting a freethinker into the Abbey was not easy" (*Charles Darwin*, 666).[24] Darwin's disaffection from religion, which grew, as I have been arguing, as his discomfort with poetry grew, derived from the same source: the facts are not consonant either with poetry or religion, and it is positively immoral to lean on untruths to evade or deny the pain of life in the interests of either aesthetic or spiritual pleasure. Darwin wanted to love poetry as he wanted to be religious, and indeed his outward public life remained pious and conformist: he was a true friend of the church in Down. But honest engagement with his own feelings made it impossible to figure out how to derive pleasure from poetry, that relentlessly human and imaginative and—in its way, despite Darwin's quest for the facts—truth-speaking mode, or how to achieve faith when he saw the world and his own life brutally and mindlessly afflicted. He could not find a way. One can argue, finally, that he belongs at last in the Abbey precisely because he resisted both religion and poetry on moral grounds, in the name of the truth, and in defense of the human sensibilities he valued above both.

So art for Darwin became good only to escape from the resonant implications of his own science—which, ironically, kept him in deeply attentive and, one must say, loving contact with the world of nature that had personally attacked him. In the passage in the *Autobiography* in which he confesses to being nauseated by Shakespeare, Darwin points out that there is some literature he likes, literature that he doesn't have to take seriously. "Novels," he says,

> Which are works of the imagination, though not of a very high order, have been for years a wonderful relief and pleasure to me, and I often bless all novelists. A surprising number have been read aloud to me, and I like all if moderately good, and if they do not end unhappily—against which a law ought to be passed. A Novel, according to my taste, does not come into the first class unless it contains some person whom one can thoroughly love, and if it be a pretty woman all the better. (138)

Nobody, alas, passed a law against Shakespeare, who, brooding on Cordelia's death, provided no escape at all. In a way, it might be said that he abandoned poetry not because it allowed escape from the truth but because in its way it was too truthful.

Science itself, the sheer spectacle of a world overwhelmingly complicated, "entangled," and wonderful beyond human imagination, became—along with the novels he could not believe—the single real consolation in Darwin's world. There was grandeur in this view of things. The wonder of particular details left him happiest when he was counting seeds in a pot, playing the piano for worms, germinating what he could find in bird droppings saturated with salt water. Out of such potentially messy, even sordid, matter emerges the sublime, emerges new life. On May 13, twenty days after Annie's death, Horace Darwin, the last of the Darwin children, was born. Out of death, life. That is Darwin's new myth of origins, echoing Milton's but without a redeeming God. Annie would not come back. Deep into his barnacles, Darwin found, not compensation and redemption, but new life.[25] Beyond the cruel indifference, beyond the death of his daughter, there was indeed new life in a world almost infinitely complex, teeming, diverse, enchanted, after all.

CHAPTER 6

"And if it be a pretty woman all the better"
Darwin and Sexual Selection

In the long argument of this book, my primary objective is to demonstrate, through the example of Darwin and of his writing, the compatibility between an enchantment that has the power to stimulate ethical engagement and a naturalistic vision of the world. Using Darwin as a model, I tried in the last chapter to begin that demonstration, noting in particular Darwin's remarkable attention to minutiae, in both his science and his life. The science and the life are entwined in a way that—in spite of the strong tradition that self-consciously splits science off from ordinary life, purifies it, objectifies it—is fairly common among scientists. Of course, it is characteristic of Darwin, and I want in this chapter to come in another way at a sense of the integration of his life and his science so important to my argument. Having focused on details of Darwin's life to demonstrate how it is integral with his science, in this chapter, I want to look at the way he formulated one of his most important theories, the theory of sexual selection, in order to suggest how integral to its construction were the conditions of his life and of his culture. That is, I want to work against the prejudice that assumes that a scientist's cultural assumptions must be kept entirely out of his work and that sees evidence of their presence in scientific thinking as reason to regard the science as suspect. Feeling and valuing are never far from objective and disinterested science, and feeling and valuing are inevitably tied closely to the culture in which the scientist, willy nilly, is immersed.

Certainly, the distance between Darwin's nonscientific life and his science was very small. The two, in fact, overlapped at almost every juncture. We have seen that part of the way he dealt with the nightmare of his daughter's death was to find a way to record it (and then her life) with the kind of detail that at least provided the illusion of some sort of control. Annie's death, in addition, recalled the problems of inbreeding, with which his work was to be often concerned. As Darwin developed his theory, he inevitably crossed the boundary between scientific, dispassionate observation of nature and direct and concerned engagement with the human: the theory of evolution, even when it worries about barnacles and plants, is almost inevitably self-reflexive.

The fact of the proximity of the human to other organisms was at the center of Darwin's entire project, and thus his anthropomorphism, which can be so surprising to modern readers who share the modern tendency to reject such rhetorical strategies as unscientific, was in fact a critical aspect of his science. "All living things have much in common," he announced near the end of the *Origin* (484), and for Darwin that sense that all living beings constituted a commonalty fed all his observations and allowed him to make quite original conjectures (hypotheses) that could guide his research and yield new ideas. Built in to Darwin's theory is a fundamental assumption (let's call it a hypothesis to make it feel more scientific) that humans are connected in important ways with lower animals. To understand those creatures, it is useful to consider what human reactions to their conditions might be. On the one hand, this might lead to a disenchanting version of his theory: spirit is driven out of the world; rationality triumphs and reveals our derivation not from a god but from some "hairy quadruped." On the other, it can lead (as it clearly did for Darwin) to a thoroughly enchanting vision of the world in which everything is infused with a spirit recognizable to human consciousness. Everything is connected to everything else.

Following Darwin in his anthropomorphic modes, the idea of a nontheistic enchantment begins to make sense. It is not some god that gives the world its meaning but the intelligence that humans share, in varying ways, with all living creatures: the worm pulling leaves into its hole by the narrow end or responding to certain musical notes; the bees that manage to create their hives in the geometrically perfect form for efficiency and storage; the ants whose colonies operate almost like a brain. Darwin repeatedly affirmed that recognizing his connection with the "lower animals" never bothered him, never diminished his sense of human value. In fact, it gave him a rich sense of what I want to call the sacred nature of nature itself—sacred as we would want to think human life is sacred.

With all that, it is important to remind ourselves—in order to avoid falling into a sentimentalizing vision of what Darwin did and an excessively cleaned up notion of nature—that Darwin's anthropomorphism had less attractive consequences as well. Anthropomorphism is not a general condition but one that is tightly located in time and place. What does it mean to see organisms as thinking and feeling like humans? It depends, of course, on one's own sense of what a human is, and this is largely determined by the way humanity is imagined at the moment and in the place in which we live. For Darwin, then, anthropomorphism would have to have meant seeing other organisms as rather like Victorian gentlemen, and this makes the anthropomorphic way of seeing seem all the more dangerous, particularly to people like us who believe themselves beyond the limits of Victorian gentlemen and their cultural prejudices. I want here to follow out Darwin's thinking as it led him through his theory of sexual selection, to try to identify the degree to which anthropomorphism limited and distorted his ideas, and the degree to which it became a means to rich and original speculation about complex biological issues.

Among the most culturally contentious of Darwinian theories, sexual selection is so ideologically fraught that it is virtually impossible to think about apart from the social imperatives it is often believed to impose. Darwin's views on the relations between the sexes appear to many to be dangerously retrograde, and one can find both in his personal writings and in his "science" strong evidence of the Victorian sexism that modern cultural critics take for granted and, of course, reject. His theory seems to do a jig and a dance that finally allows him to affirm the physical and mental superiority of men to women. Despite the fact that many nineteenth-century feminists thought of themselves as Darwinian, feminists now tend to look at Darwin as dangerously and influentially chauvinist, naturalizing Victorian social conventions and thus giving scientific sanction to his culture's prejudices about women.

In the same sort of language that Darwin uses so effectively in describing the physical characteristics of insects, birds, and mammals, he writes, for example, that the formation of the woman's skull "is said to be intermediate between the child and the man" (*Descent*, 2:317). He cautiously rejects the view that women and men are equally intelligent. But the difference, he says, is made probable by analogy with lower animals: "No one will dispute that the bull differs in disposition from the cow, the wild-boar from the sow, the stallion from the mare, and, as is well known to the keepers of menageries, the males of the larger apes from the females" (2:326). And he then goes on with a set of Victorian truisms about women's greater "tenderness and less selfishness" because of their "maternal instincts," their greater "powers of intuition, of rapid perception, and perhaps of imitation." These are the usual Victorian sops to women, who, Darwin suggests, are otherwise inferior to men in "deep thought, reason, or imagination, or merely the use of the senses and hands." So, at last, he buys into John Galton's arguments that "the standard of mental power in men must be above that of women" (2:327). One doesn't have to shop around very far in the vast literature of Darwin and Darwinism to find evidence that the accusation that Darwin's theory bears all the marks of Victorian cultural prejudices is accurate.

Accurate but inadequate. There is no question that part of the way Darwin succeeded in making evolution attractive to his contemporaries was through his sometimes unreflective sharing of many of his peers' cultural prejudices. Certainly, he was no firebrand revolutionary. Anything but. We have already seen how the recent biography by Desmond and Moore, and Janet Browne's even more detailed study, show Darwin comfortably surrounded by a network of politically and socially alert defenders who, while they might use Darwin to take a jab or two at the clergy, were very deeply committed to hierarchy, and certainly largely convinced of what women's place should be. But it is worth looking closely at some of what he actually said about female insects, birds, mammals, and women in his development of the theory of sexual selection. His writing about it is anything but univocal, and it provides a fascinating example of the way in which cultural assumptions operate in his theorizing and were fruitful in allowing him to think around conventions. I am not going to try to suggest what the political implications of Darwin's arguments *should* be, but I do want to trouble at least one aspect of what they have often been. And I want to do this within the larger argument of this book to demonstrate in another way how Darwin's relentless attention to minutiae and his feeling for organisms opened up an enchanted world that, at the same time, became scientifically intelligible.

The conventional view is that conventional views, like those Darwin held about women, damage independent and serious scientific research; either that, or they infuse themselves into science so thoroughly that it is absurd to think of science and culture as radically divided and divisible. Looking at what Darwin actually achieved, I want to argue that his very conventionality helped him to think through positions that could, in the end, subvert the conventions that encouraged them. Darwin's science was surely much influenced by his social and cultural position, but this neither invalidates the science nor implies that what he did was always constrained by those cultural limits. Here is a place where Darwin's work, like most interesting literary and scientific texts, outstrips the cultural limits that allowed for their very production. If we are all condemned to say only what our culture has already said, there could be no new ideas and no new knowledge. As a commentator on an early version of this chapter put it, "Darwin's ideological assumptions about gender . . . helped generate a theory that has proved to be, in its central points, correct; and . . . to some extent subversive of the very assumptions on which it was based."[1]

As is well known, Darwin was always, even in and after *The Descent of Man*, very cautious about the degree to which his ideas about evolution actually applied to man. In the long run, certainly, he believed that one major test of his theory, perhaps *the* major test, would be the degree to which his ideas actually applied to the human. So he had a large stake in his argument that "It is . . . highly probable that with mankind the intellectual faculties have been gradually perfected through natural selection" (*Descent*, 1:161). The difficulty, he understood—greater even than in his argument for the development of the eye through natural selection—would be to make the case that those elements of human life that seem to be most distinctive, that most differentiate it from that of the lower animals, are the effect of merely natural causes, and the same sorts of causes that went to the development of apes and tigers. Darwin wanted to argue that every aspect of human nature, including the ethical and esthetic powers, could be explained (but *not* explained away) by the processes of natural selection. Out of inherited instincts that conduced to survival and reproduction emerged just that sense in us for goodness, that sense in us for beauty that Matthew Arnold lamented was so sadly lacking in Darwin's descriptions of the origin of the human species.

Even the greatest of Victorian scientists trembled at the prospect of naturalizing ethics and aesthetics. Wallace, whom Darwin in fact invokes as he begins his investigation, drew back at the prospect and claimed that even though natural selection worked on every aspect of bodily development, it could not possibly be responsible for human intelligence and the virtues of higher civilization. As Martin Fichman summarizes Wallace's position, "According to the utility principle, natural selection would have provided the savage with an intellect only slightly superior to that of the apes. It cannot, Wallace emphasized, explain the complexity of the savage's brain" (192). These and other conclusions of a similar kind led Wallace to believe that for the

full development of man, there had to have been the intervention of a "higher intelligence." It should be noted, moreover, as
we consider the political and social implications of these ideas,
that Wallace was on almost all such matters far to the left of the
modestly Whiggish and wealthy Darwin, so that here we have a
case of an incipient socialist adopting a position that one might
have thought more consistent with the political conservatism of
a Balfour or an Argyll or a Mallock.

In any case, Darwin was of course deeply disappointed by
this turn in Wallace's thought. But the sheer mindlessness of the
processes Darwin described was disturbing to virtually everyone, and Lamarckian intention became a very popular variation
on Darwinism in the late nineteenth century, particularly among
the Fabians. The need for meaning and design in the universe
was not reducible, then, to a dichotomy of right/left, or tradi-
tion/modernity.

The rigor and tenacity with which Darwin stuck to his theory
of natural selection (there was much bending but no breaking) is
one of the marvelous facts of his intellectual career. And as with
natural selection, so with sexual selection, to which only a very
few Victorians ever subscribed: Darwin withstood Wallace's powerful objections to the idea and devoted the larger part of the *Descent* to making the argument in great detail. But objections to his
sexual selection theory have survived, though in very different
forms, right through the twentieth century. From the start, for
reasons I will discuss shortly, Wallace found the theory inadequate, believing that what needed to be accounted for was not
the bright coloration of males but the drabness of females, and
that, he thought, was strictly the work of natural selection pro-
tecting the breeding or brooding female from attack. On the other
hand, as I have already suggested, many modern critics quite legitimately find the theory sexist. But whatever the critique, the
first important point is to recognize that Darwin was, as always,
trying to account in naturalistic terms for qualities that seem to
be exclusively human, without positing, as Wallace felt required
to do, intelligent intervention from outside the sphere of nature.

Perhaps the most striking aspect of Darwin's theory of sexual selection is that, as Gillian Beer has argued, it puts back into the theory of evolution the intention and cultural direction that seemed so alarmingly absent in Darwin's first formulation in the *Origin*. The absence of intelligence in evolutionary direction amounted to a cultural crisis, but the intelligence that Darwin reinserted into the process by way of sexual selection did not entail the reintroduction of some extranatural mind of the sort to which Wallace was to turn. The intention that sexual selection depends on derives from the activities of the developing organisms, not from a supervening master intelligence. It is, as it were, a contingent intelligence, but it does provide direction, under the ultimate severe control of natural selection. Darwin thus keeps the process entirely natural while in a way accounting for the sense of intention that most humans have about who they are and what humanity is all about. It is just that for Darwin, intention is simply a part of the natural process of struggling to live, struggling to procreate. Darwin's theory, then, in effect reproduces the form of previous, metaphysical explainations of origins and of nature, but without God. However it may be regarded now, it was a daring move.

While, as we have already noted, in the *Origin of Species*, Darwin was careful not to discuss the human species or culturally difficult issues (except, inevitably, through metaphor), in *The Descent of Man* he met humanity, morality, and culture head on, not only to attempt to explain them naturalistically but to introduce cultural determinants into biological developments, or vice versa. It was a striking and difficult move, one whose effect was yet more dubious and controversial than the theory of natural selection itself. Intention, that central motif of natural theology, went out with the *Origin*, only to return with the *Descent*. The disappearance of "intention" is perhaps the most striking and most disenchanting aspect of Darwin's theories, but in *The Descent of Man*, the intention that disappears from the development of the species, mankind, returns in the development of sexual difference. And, of all things, it returned primarily by way of female choice. This was the most daring and the least appreciated move of all.

It is time to recognize boldly that although Darwin was very much a man of his time in his attitudes toward women's place, he turned not to God but to females to account for the initial sense of direction and intention that life manifests. Darwin is the man who, when thinking about whether he should marry, noted that a wife would be a "constant companion . . . [an] object to be beloved and played with . . . better than a dog anyhow."[2] But he is also the man who, against the whole scientific and cultural establishment, introduced the notion of female choice into nature. There is no way to talk about sexual selection without bringing to the surface all those cultural issues that Darwin tried hard to keep back, or at least hold in abeyance, as he made the case for his primary argument about evolution and its mechanisms. But no reasonable discussion of Darwin's influence should fail to come to terms with this striking introduction of female choice and, implicitly, female intelligence in a process that Darwin argued was fundamental to the development of men and women, and of racial difference. The idea may even have been against the grain of Darwin's own commitments, and when pushed to it, he would probably have denied that women are in a position to exercise any power over the direction of evolutionary development. But there, more than latently in the *Descent*, are the females making their selections.

Sexual selection is, thus, a minefield, not only for scientists, but for cultural critics as well. Nothing is easier than to fall back on the critique that Darwin's theory is really only a naturalizing of cultural prejudices about women, as it has been claimed that natural selection is nothing but a naturalizing of capitalism. Yet where the "nothing but" in regard to natural selection dissolves into absurdity, sexual selection remains contentious enough to be threatened by its obvious relation to well-known Victorian prejudices about women, and that relation can be (or has been) taken as its virtual death knell as a serious biological explanation.

But this ideological critique, in both cases, has no purchase on the theory itself. It might indeed help us to understand the history of the development of the theory, but it cannot do much to validate or invalidate it. However embedded *The Descent of Man* might be in uninterrogated cultural assumptions, the theory of sexual selection is an astonishingly brilliant idea, teased out of a mixture of cultural assumptions, intense observation, and careful thinking. These cannot be disentangled, and their entanglement (a deeply Darwinian concept itself) is part of the reason that ideological critique cannot, by itself, dislodge the theory, though it may provide good reasons to like or dislike it. Good cultural theory might best take sexual selection not as a simple reflex of cultural prejudice but as a fascinating commentary upon it.

Unfortunately, while Gillian Beer's approach to Darwin, which follows out such lines, has happily transformed almost all later cultural study of Darwin and his relation to literature and language, treatment of the question of sexual selection has not, for the most part, profited from her approach.[3] It is one of the distinctive and most satisfying qualities of Beer's remarkable treatment of Darwin in relation to culture that she talks freshly and revealingly about the ways in which cultural (dare one say "nonscientific"?) forces operate on Darwin's thinking without feeling obliged to judge him or his theories negatively from the perspective of our current ideological positionings. Indeed, she takes such judgmental critique as a symptom of an intellectual arrogance that closes down the possibility of learning from the past. Darwin is troublesome and troubling, never moreso than as Beer has read and used him critically. One learns from her about how Darwin's thought developed in intimate contact with literary and cultural forces that are normally ignored in strictly scientific discussion, and about how his thought percolated through the culture—but in every case with a sense of the instability and richness of his language, and never with a sense that as a wealthy male citizen of his own time, he can be satisfactorily understood as representatively retrograde ideologically. She simply will not accept the all too common assumption that Darwin's implication in the values and ideals of his own culture somehow closed off the possibility that his work could extend beyond the limits of that culture to make genuine discoveries and to criticize it. As she put it, Darwin's "writing intensified and unsettled long-used themes and turned them into new problems."[4]

It is worth, then, following Darwin's theory into the very heart of cultural complicity, where nature and nurture, culture and science, get hopelessly entangled, and where we might begin to get a richer sense of its originality and difficulty. As I do so, I feel the pressures of conflicting allegiances. In my studies of Darwin I have been much impressed by, and have learned much from, the sort of criticism I have discussed in the opening chapters, criticism that has demonstrated the connections between Darwin's theories and his ideological assumptions, about politics and gender in particular. But I have also always found useful and important the approach to Darwin that reads him very closely, follows out the historical lines of the ideas he uses and develops, and emphasizes both the ways in which he developed his theory and the nature and originality of his arguments. Such approaches ought not to be incompatible. The first primarily emphasizes Darwin's ideological complicity, so that his work might begin to look like little more than an elaborate apology for some very bad Victorian habits—the path from Darwin to eugenics or libertarianism is then direct! The other approach leans toward hagiography, the abstracted celebration of original genius, that I have been trying so hard to avert even as I perform some of the work of hagiographers.

While good cultural criticism does not reduce science to ideology, good intellectual history does not leave ideas and genius hanging abstractly out there without context. The extraordinary historical work of Desmond and Moore—which often does, indeed, combine very close examination of language with broad contextual and ideological placing—is perhaps the best example of historical and cultural criticism that is ideologically driven but conscientiously committed to getting Darwin right, understanding how he thought, and watching both the cultural processes that gave shape to some of his ideas and the unique intellectual and personal qualities that allowed him to produce his theory as he maneuvered through social conditions in which he felt extremely comfortable but whose very foundations his theory threatened to challenge.

Desmond and Moore argue that the theory of natural selection locked into place as Darwin found a form for it compatible with the economics of laissez faire, and strong historical study like theirs seems to me to have established once and for all that the theory has close and documentable ties to laissez-faire economics.[5] Desmond had previously demonstrated, in *The Politics of Evolution*, that early-nineteenth-century evolutionism had largely radical and even revolutionary roots, and no doubt Darwin had to work hard to disentangle his version of the theory from what he would have regarded as politically tainted ones. But, as I have suggested, the biggest political bang of natural selection came from the evacuation of intelligence and direction in the development of species.

It was never necessary for Darwin to announce this absence, but only to follow out the implications of the remarkably simple and overwhelmingly complicated chart that he set into the chapter on natural selection (as he talks about "divergence of character") and to which he reverts with more elaborate variations several times in the course of the *Origin*. I have already given an example of a modern use of this sort of argument, Ridley's on Balinese rice farming, although Ridley's is overtly linked to a political program. Nevertheless, the "Leviathan" that Ridley laments is just the intelligent god that Darwin excludes. Nature does its work without the intervention of either. There is an ostensible randomness about Darwin's chart, however, and that randomness is itself a reminder of the absence of a shaping intelligence. Things, on the chart, just shake out to produce new species. The chart is, however, itself an intelligent intervention in that it represents a brilliant thought experiment, in which species do not develop according to some blueprint but spin out of each other, fail, or just plow straight ahead in a diagram that has a kind of symmetry but is most distinctly more like a tree than a geometrical figure. Intelligence and direction inhere only in the experimenter, not in the work of speciation.

There was, however, no way to leave either intelligence or politics out of sexual selection. Here our contemporary critics go straight for the culture, and it will be useful to look at a pair of important essays on the subject as a way into an understanding of the relation of the culture to the theory, and of the theory's power to be larger than its own politics.

Rosemary Jann's strong and tightly argued essay on sexual selection remains a point of departure for any discussion of the subject, and in my attempt to complicate matters, I take it as my point of departure as well.[6] I would add to it an important essay of about twenty-five years ago by Eveleen Richards, who claimed in what was then a groundbreaking argument, that "Darwin's re/construction of human evolution was pervaded by Victorian sexist ideology."[7]

Those essays have been important in demonstrating that Darwin's very unscientific assumptions about the place of women played a role in the development of the theory of sexual selection in *The Descent of Man*. There is no need to follow the details of their strong arguments, but surely it is difficult now to imagine picking up *The Descent of Man* without being struck by its thorough saturation in cultural assumptions. Darwin's anthropomorphizing is a central feature of his work, and virtually every one of his animals and insects seems to behave in very Victorian ways. Much of the *Descent* is anecdotal. Right at the start, Darwin confesses that "This work contains hardly any original facts in regard to man" (*Descent*, 1:3), and the absence of originality in these factual matters is striking. For the most part, Darwin summarizes his earlier notes and the views of others, often without the extraordinary care and rigor that mark his arguments in the *Origin*. Yet care and rigor do mark the arguments of the largest parts of the book, the sections dealing with sexual selection, mostly of plants and animals. It is when he gets to talking about people that Darwin turns, it would seem almost defensively, to the views of others (he notes on the first page that he didn't talk about humans in the *Origin* because he feared doing so would "add to the prejudices against my views" [1]). The *Descent* is often disappointing in parts, even to Darwin enthusiasts like me, because so much of its discussion of human behavior seems to depend on commentaries by second-rate minds on Darwin's earlier first-rate work. Much of it relies on the materials provided by contemporary ethnologists and anthropologists, who were themselves ideologically saturated at a moment when the new social sciences were being born. Even in the *Descent* Darwin's most interesting ideas about the human tend to get there only by indirection. His discussion of plants and animals, meticulously derived from close observation and carefully accumulated work, is saturated with his sense of the human, and what emerges is an extraordinary mental tour de force.

In arguing this, I do not mean to suggest that the Darwin whose personality gets expressed in the language of this book is anything but a respectable middle-class gentleman (one who, by the way, never allowed himself to be pictured, as Janet Browne has shown in "I Could Have Retched All Night," as a man marked by his work in any way). The cultural critiques by Richards and Jann were necessary and largely right. The respectable middle-class gentleman emerges from just about everything Darwin wrote. There is the homeowner who puttered about in his garden, who modestly took into account all possible objections to his ideas, who steered clear of controversy even while creating one, and who treated respectfully all who helped and all who disagreed.

But what do the arguments of Richards and Jann mean for the *validity* of the theory of sexual selection? What might Jann have meant when, near the end of her essay, she wrote: "Acknowledging the extent to which the imposition of order on events is necessarily dependent on the ideological position of the observer who interprets data and fashions stories from them need not rule out the possibility of satisfying scientific standards of proof and logic in our reconstructions of the past" (304). A remark in Richards's essay raises a similar question: "Darwin's conclusions on the biological and social evolution of women," she says, "were as much constrained by his commitment to a naturalistic or scientific explanation of human mental and moral characteristics as they were by his socially derived assumptions of the innate inferiority and domesticity of women" (61). That sentence seems rather oddly phrased. On the one hand, it suggests surprise that scientific explanation might account for anything in Darwin's scientific theory. On the other, it suggests that the focus of Richards's arguments will be where it is not, that is, on the demonstration of the "scientific constraints." Nevertheless, it is extremely interesting and important that in two excellent essays implying the primacy of cultural explanation, both writers insist on the necessity of scientific constraints. The question that arises from that insistence, however, is what can be meant by such constraints and how such constraints can even be imagined. Are the authors suggesting that there are some issues that are constrained by the intrinsic nature of their materials *as opposed to* constraint by cultural forces? This, of course, is the fundamental argument of those who have taken up the cudgels for science in the recent lamentable "science wars." There are some facts (which scientists tend to discover) that are simply there regardless of our point of view, and that emerge in any serious study, whatever cultural constraints might also be operating. But beyond what I take to be an oversimple division between scientific and cultural restraints, one might ask the question, what if, as some cultural historians insist, *everything* is culturally constrained in some way or other? What might be the consequences of this assumption for the study of particular scientific arguments? What might this mean for the theory of sexual selection, in particular?

That is to say, after all the evidence is in about the cultural forces that shaped Darwin's arguments, there remains the question of whether Darwin was right about sexual selection itself. "Sexual selection," writes Fiona Erskine, "is intrinsically antifeminist."[8] There are two questions to ask about such a claim:

does it also imply a claim about the *validity* of the theory? And can sexual selection in all of its complications be contained by such an ideological implication? No doubt, when Darwin came to write about humans he largely reinforced, as Erskine says, his deeply rooted and widely shared views about the inferiority of women (though he had long supported women's education). But subtle critical analysis like Jann's forces an unpacking of the question, for her essay locates contradictions within Darwin's own arguments that seem to make it impossible—except through the wildest chance—that they were entirely right (and it is widely agreed now that they were *not* entirely right). But there remains to this day a likelihood that Darwin's fundamental ideas about sexual selection among animals were correct, and it would be a very bad mistake to dismiss the theory solely because Darwin's sexism emerges so clearly when he talks about humans. In fact, to do so would be in effect to leave oneself open to the possibility that it is not only Darwin but nature that is intrinsically sexist. We are past the stage, either in cultural studies or in science, of having to take Darwin whole cloth. We should also be past the stage of thinking that an argument ends once it is demonstrated that a position is ideologically constructed. If it is universally the case (and the need to invoke the word "universal" in such a matter is an ironic commentary on the futility of the "everything is political" argument) that everything is ideological, then the important work of analysis and understanding must begin *after* the ideological work is recognized. Nothing is proved by proving that a thing is "ideological."

This is not the place, nor is there world enough and time, to discuss what it might mean to hold to scientific standards of proof and logic when we also believe that every discourse, including our own, must be marked by local cultural perspective. While I believe that there are legitimate standards of proof—though perhaps not so systematically ordered and recognized as is sometimes claimed—that resist the pressures of cultural forces, my argument about Darwin's peculiar genius here depends on an at least provisional acceptance of the general view that his ideas about sexual selection (and even natural selection) were significantly informed by his cultural assumptions. Obviously, on the understanding that cultural perspectives are pervasive, an answer to the question of what constitutes scientific standards of proof and logic would inevitably be determined by those very perspectives.

However such views are interpreted, some recognition of constraint in interpretation is as necessary to commentators on Darwin as Darwin felt them to be on himself. Darwin was certainly constrained by other things besides his ideological assumptions. He was alert to the possibility of the kind of criticism that cultural critics have been leveling at him as when, near the end of *The Descent of Man*, he plainly asserts that "The views here advanced, on the part which sexual selection has played in the history of man, want scientific precision" (2:383). The question of canons of scientific validity was serious for Darwin. He knew the degree to which his various rich arguments about natural and sexual selection required much more substantiation before they could be firmly established, particularly with regard to their roles in the development of humans. He worried, legitimately, and not simply in the throes of ideological blindness, about whether he could produce enough evidence, and his whole work is marked by rhetorical admissions that if such and such were to be the case it would be fatal to his theory. He certainly knew that his discussion of humans was yet more speculative than any other part of his work. But surely, having read Darwin's work and his letters and notebooks, the most hardened critic would have to concede that his highest priority was not to enforce his sexist assumptions or his preferred economic theories but to get it right. His passion for the world and for his study of it is manifest again in the attention to minutiae that marks every step of his argument about sexual selection. In his own speculativeness, of course, he believed what he argued while at the same time recognizing the vulnerability of his position.

On the subject of sexual selection, Darwin knew in how small a minority he was. But if he was being ideologically complicit in insisting on his theory, what about the complicities of those who disagreed with him? On the one hand, there was A. R. Wallace, more Darwinian than Darwin, though politically much to Darwin's left. On the other, there was St. George Mivart, the strongly Roman Catholic scientist, whose religious and political positions were obviously to the right of Darwin's and who talked of "vicious feminine caprice," which made the idea of female choice helping establish permanent evolutionary changes absurd to him.[9] We might, of course, rub out the ideological differences among those who rejected Darwin by claiming that on at least one issue they were all in it together, trapped in the culture's prejudices about gender, but such erasure makes important distinctions impossible and suggests that nobody at the time could have either opposed or supported Darwin's theory without being guilty of the same sin. Hardly an intellectually profitable argument.

One of the more interesting questions for the cultural study of science, then, would be whether rejection of Darwin's theory meant rejection of the culturally pervasive ideological assumptions now attributed to it (which would be odd). But if it did not mean that, what would that conclusion suggest about the direct relation between Darwin's theory and any particular ideological assumptions? Particularities are what would be needed in discussion of the various positions adopted by those opposed to Darwin, and what would it signify that those various positions all implied the *same* cultural assumptions? I hope I am not being merely naïve in claiming that in the unlikely event, at that historical moment, that Darwin had been presented with evidence that led to the conclusion that women were not inferior to men, he would have accepted it as "fatal to his theory." Part of what it meant for Darwin to be in love with his science and with the world is that (within, of course, the possibilities that his own cultural limits imposed) he listened to and could hear what nature was saying to him. He gave to nature the full sympathetic imagination that could allow him to find those places where nature's voice was different from what his culture might have been telling him. Such a concession could not have been greater for him than the one his discoveries of sexual selection forced upon him, that there was some intelligence driving the motors of evolutionary development after all. The way Darwin's theory disrupted the natural-theological understanding of organisms entailed of course the rejection of the notion that they were intentionally designed and directly adapted to their positions in the world. Recall that Darwin makes his claims about the influence of female choice on evolutionary development in the same book in which he confesses to having mistakenly assumed in his earlier work that "every detail of structure, excepting rudiments, was of some special, though unrecognised, service" (1:153). *The Descent of Man* presents a theory of evolutionary change that is self-consciously antagonistic to the idea that everything has a purpose, that all "details of structure" are of use. Sexual selection, however, is a theory that depends on the assumption that only by recognizing that dimorphic details are indeed of some "service," after all, can one make sense of racial and sexual difference.

It is fascinating to consider Wallace's relation to the theory, if only because Wallace's work was so distinctly entangled in other political and social ideas. Wallace was not only opposed to Darwin's theory of sexual selection but was a much more rigorous adaptationist than Darwin. He believed that (putting aside the mind of man), virtually all details of structure in organisms were of "use," that the uses were adaptive by way of natural selection. (And it would be fascinating, if somewhat beside the point, to notice how askew from consistent political positioning the debates about these issues are. In the most recent battles, Stephen Jay Gould, whose opposition to the politics and theory of sociobiology was so frequently expressed, has been the anti-adaptationist, arguing that many characteristics of organisms cannot be explained by the working of natural selection or are mere accidents of its processes. Wallace then was politically closer to what Gould is now than the antiadaptationist Darwin would be. But part of the point of my making this parenthetical comparison is to argue against the idea that any of these theories somehow occupies a necessary position on a political spectrum.) Wallace's *Darwinism* was, as Martin Fichman has argued, intended in part "to demonstrate that the varied phenomena of sexual dimorphism could be subsumed under the action of natural selection" (267). Wallace simply couldn't believe that the aesthetic sense in animals or humans was dependent on natural evolutionary forces. Darwin's hard-nosed naturalism made the brilliant and politically energetic Wallace very uneasy, though his own commitment to natural selection, as it worked in animals and in man (except for his intelligence and spiritual capacities), was even stronger than Darwin's. I wonder what would have been the fate of evolutionary theory in relation to culture and politics if history had allowed the priority to Wallace instead of Darwin. (And parenthetically, it might also be noted that on the crucial question of human intelligence Wallace lapsed back, as it were, from problem to mystery, in the sense that the only answer he could find to the "problem" was what might now be called an "intelligent designer." Here is the space for a renewal of enchantment, although how it would sit against a physical evolution entirely driven by natural selection is hard to imagine.)

Whatever Darwin's own views about intention, any discussion of the ideological implications of his theory of sexual selection ought to take account of perhaps the most striking fact about it: it was not only Wallace who did not believe it; virtually nobody did. To be sure, there is Grant Allen, a very strong Darwinian, whose interesting and extravagant book, *Physiological Aesthetics*, follows out Darwin's view that aesthetics was born from sexual desire. There is also strong evidence that Hardy read and was influenced by *The Descent of Man*. But any consideration of the cultural influence of the idea of sexual selection must take into account the fact that within biological science sexual selection had no serious place for many years—not until, in fact, quite recently. As Helena Cronin has shown in her extensive survey of Darwin's theories, "There has been very little discussion of sexual selection through the years and it is only now being taken seriously historically and scientifically"(115). But Darwin's importance assured that the theory would generate some discussion of the possible role of sexual relations in evolutionary development. Unlike natural selection, which was also not scientifically accepted until well into the twentieth century, sexual selection does not seem to have seeped into the culture very deeply, though sexism, scientific and otherwise, was pervasive. Until very recently it was possible to treat the theory as Gertrude Himmelfarb has done, as public evidence not only of Darwin's recognition that his theory of natural selection had failed but also of his intellectual shallowness.[10]

That the ideological work Jann and Richards detect was actually going on, I do not doubt. But that Darwin's theory of sexual selection had much influence on Victorian culture is unlikely. Cynthia Russett claims that although the view that the male is more variable than the female was indeed used to do some dirty ideological work, that idea did not at all depend on Darwin's theory for its support. She argues that "The elaborate edifice of female conservatism and male progressivism, female mediocrity and male genius, that was presently erected on the foundations of variability, does not derive immediately from *The Descent of Man*."[11] It is, of course, no defense of Darwin's argument, one way or the other, to say that it was not influential, but it is nevertheless noteworthy that it had very little immediate influence and largely because the scientific community found it impossible to credit the idea that the female could have had much to do with evolutionary development.

To elaborate yet further one of the dominant motifs of this chapter and of the book as a whole, I want to refer here briefly to the arguments of Oscar Kenshur, who has labeled the view that particular scientific theories have particular ideological implications "ideological essentialism." He makes the point this way: "the ideological essentialist wants to be able to find intrinsic political significance in abstract theories that at first glance seem lofty and disinterested."[12] The alternative to this view, the one I have already urged, Kenshur calls "ideological contextualism" which I regard as the appropriate response to "essentialism." In offering a few of the "uses of Darwin" for inspection in the second chapter, I have tried to make clear that not only Darwin's theory as he developed it, but the theories of those who subsequently used it were developed within contingent conditions that inflected his thought and that made it polyvalent. That is, the theory could be (and continues to be) put to uses that probably have little to do with the uses Darwin, working in his own contingent world, might have imagined for it.[13] What remains after the uses and the misuses is always the theory again, to be appropriated and misappropriated. Moreover, as I have tried to intimate through my brief allusions to Wallace's views on sexual selection, it is simply too difficult to disentangle the threads of argument enough even to define precisely what the point was that led to an inevitable political implication.

The basic theory of sexual selection, although after 150 years of ideological exposures it may be difficult to recognize, remains fertile and disruptive. Its brilliance and originality are, as Darwin's language makes clear, intimately connected with its cultural sources. Having identified the gender prejudices of the culture that play into Darwin's imagination of sexual selection, one will find that the theory itself forces a break with just those prejudices that produced it, and Darwin's reversal of his argument from animals to humans is a particularly good sign that his thought outleaped the culture that helped form it. (To further confuse matters, Wallace too reversed himself on the matter of human female choice and ended by accepting sexual selection after all, at the human level. His reversal clearly reflected, finally, his political views of woman's place in culture, just as Darwin's reversal at the level of the human reflected his.)

But how could Darwin have come up with the idea that the aesthetic sense derives from animal sexuality? The difficulty of the conception is multiplied when one understands that it required that he recognize that female choice was at the root of it all, despite the fact that Wallace didn't buy it, that, as Russett points out, virtually nobody did. Cronin describes a representative position, taken by Darwin's opponent, St. George Mivart, which claimed that female birds simply could not have the sensibility to respond to the aesthetic appeals of ornamented males. So, he claimed, "the female does not select; yet the display of the male may be useful in supplying the necessary degree of stimulation to her nervous system" (157). Females are, after all, very coy. And the whole idea ran counter to Darwin's own instincts on the matter. Frederick Burkhardt quotes Darwin, in a passage written shortly before he died, as registering that "many naturalists doubt that female animals ever exert any choice so as to select certain males in preference to others. It would, however, be more correct to speak of the females as being excited or attracted in an especial degree by the appearance, voice, &c. of certain males rather than of deliberately selecting them."[14]

Mivart's cultural prejudices against Darwin's arguments were, then, probably shared by Darwin himself. But I want to emphasize that Darwin's idea of female choice derived as directly from his own unarticulated cultural assumptions as did his transference of choice to the males in humans. Darwin's working through of his theory suggests that discovery of the presence of cultural assumptions at work in scientific arguments need not undermine those arguments; in Darwin's case, at least, the presence of cultural assumptions made possible some good and innovative science, having the potential for implications counter to the very assumptions on which he unself-consciously drew.

Notoriously, Darwin did not work on the "Baconian principles" he claimed to use. The "hypothetico-deductive" method, as it has been called admiringly by interesting commentators like Michael Ghiselin, [15] begins with something like a guess. The "guess" in the *Origin* that family resemblances among species are literal was suffused with cultural assumptions. Darwin's science *needed* the ideological assumptions that cultural critics and scientists alike, for different reasons and in different ways, regard as evidence of bad science. The hypothesis, the guess, that underlies sexual selection not only indicates the way in which Darwin's ideas participate in the culture's ideologies but suggests that cultural assumptions are inevitable for anyone not in the dreamed of "nowhere" of absolute objectivity and universality. That is to say, scientific ideas, those elements that Weber says disenchant the world, are often—particularly when they have to do with the study of the human—saturated with the world and with the feelings and values that come with it.

My predisposition toward both of the two ostensibly opposed approaches I have mentioned has led me to try to understand, in Darwin's case at least, not so much how cultural prejudices and assumptions expose sexism or imperialism or other modes of complicity, but how they managed also to be creative, to produce a theory like sexual selection. Here is a quick glance at an aspect of Darwin's sexism.

Those novels that Darwin professed to think first rate, the ones that contain "some person whom one can thoroughly love, and if it be a pretty woman all the better" (*Autobiography*, 138–39), can offer a valuable clue to the interaction of culture and science in Darwin's work. In that passage, which began with a lamentation about his loss of aesthetic sensibility, Darwin rather unselfconsciously falls back on the source of the aesthetic, which he had identified in *The Descent of Man:* the "pretty woman." In the *Descent* he had argued that the "sense of beauty" was not "peculiar to man" (1:63). Careful to point out that the sense of beauty is not single, he had shown himself aware that "high tastes," as he calls them, depend "on culture and complex associations" (1:64). But what he did not quite call lower tastes are also fundamental to complex animals, and they grow from regard for the "pretty." The assumptions of Victorian culture are obviously at work here, and Darwin includes in his argument in the *Descent* some rather unpleasant (retrospectively) material about the inferiority of "savages," whose aesthetic tastes are often, from his point of view, inferior to that of birds: "Judging from the hideous ornaments and the equally hideous music admired by most savages, it might be urged that their aesthetic faculty was not so highly developed as in certain animals, for instance, in birds" (1:64). In

the context of his autobiographical comments, however, when he
self-mockingly describes his love of happy endings and pretty
heroines, Darwin implies that he has fallen back onto a primitive
sense of the beautiful—a sense, by the way, that in the *Descent*, he
had shown to be common in "barbarous tribes" where female
choice was still the norm. Pleasure in the novel, itself largely recognized as a feminine form, was surely an indication of a lapse
away from high culture back toward primitive feeling, and the
attraction of "pretty" women marks an appeal to fundamentally
primitive desires. Higher culture and a higher level of evolutionary development shifted the power of choice to men, but in the
Autobiography Darwin sadly concedes his own fall from that
higher culture.

The "pretty" attracts him as it attracts the birds, and he does
not try to explain it. In the *Descent* he confesses: "Why certain
bright colours and certain sounds should excite pleasure, when
in harmony, cannot, I presume, be explained any more than why
certain flavours and scents are agreeable; but assuredly the same
colours and the same sounds are admired by us and by many of
the lower animals" (1:64). Beer points out how, in his discussion
of the human female, Darwin steadily "gives primacy to
beauty" over intelligence. Beauty is the key concept, one that en-
tices him throughout and one that is obviously conditioned by
Victorian expectations.

Darwin's continued enthusiasm for pretty girls in novels can
serve as a reminder of how he arrived at the theory of sexual selection. The pretty girls of Darwin's unreflective, lower pleasures
are a condition for his theory. As he puts it succinctly in the *Origin*, "when the males and females of any animal have the same
general habits of life, but differ in structure, colour, or ornament,
such differences have been mainly caused by sexual selection"
(89). What makes this insight possible is Darwin's assumption
that those differences in appearance are noticeable and attractive
to the opposite sex. Not only that, but once these differences are
noticed, the opposite sex can make a choice about them, just as
Victorian gentlemen (and maybe an occasional country girl) do.

One possible inference from this obvious reliance on cultural prejudices is that there must be something wrong with the whole theory. Another, the one I want to make here, is that Darwin's experience and sharing of those Victorian tastes made clear to him a problem and suggested a resolution that is really entirely anthropomorphic and at the same time almost certainly right. The flood of metaphors that does anthropomorphic work in the development of his theory reveals the degree to which his assumptions about human culture helped shape his scientific arguments. But the anthropomorphism throughout Darwin's work, as I have already suggested, is also a consequence of his view that all organisms are quite literally related. The obviousness of Darwin's assumptions about Victorian prettiness threatens to obscure the fact that there is nothing inevitable about the theory of sexual selection that he derived from them. Even those most sympathetic to Darwin's theories to this day have trouble distinguishing between the effects of sexual and those of natural selection. Characteristics developed by sexual selection have no consequences in natural selection; they are not conditions for the survival of the organisms. Indeed, if the theory of sexual selection is correct, certain characteristics necessary for sexual selection are dangerous in the world of natural selection—the most obvious, of course, being the often striking colors and displays of breeding males. (On the other hand, the development of weaponlike characteristics, like pointy horns or deadly claws, might easily be interpreted, as Wallace tended to do, as elements in the work of natural selection.)

The theory is at times yet more counterintuitive than natural selection was. Victory in sexual selection depends, as Darwin says, "not on the general vigour" of the male but on its "having special weapons" (*Origin*, 88). Those weapons are not matters of life and death but are necessary in the struggle with other males of the same species to win the female and produce the most progeny. Since most females are not as ornate as most males and yet survive in nature as well as the males, it follows that the ornamentation has another purpose, and Darwin inferred that the purpose was to win the female. What else can all that finery be for?

In the *Descent*, Darwin unembarrassedly tries a thought experiment, a device he used brilliantly in the *Origin*, and, it would seem inevitably, it spins around a pretty girl:

> With respect to female birds feeling a preference for particular males, we must bear in mind that we can judge of choice being exerted, only by placing ourselves in imagination in the same position. If an inhabitant of another planet were to behold a number of young rustics at a fair, courting and quarrelling over a pretty girl, like birds at one of their places of assemblage, he would be able to infer that she had the power of choice only by observing the eagerness of the wooers to please her, and to display their finery. (2:122)

Here (a point I will be developing further in the next chapter) the scientific enterprise depends upon a feeling for the organism, an act of imaginative sympathy of the sort George Eliot called for in her novels. The world so perceived is thick with feeling and value as Darwin enters into the "mind" of the female birds. It is hardly that dead, arid place that Weber tells us is the consequence of scientific explanation. Much of the *Descent* depends on placing *ourselves* imaginatively in the condition of some other being. When we do, even as we try to imagine otherness, the other gets to be rather like us, rather Victorian. The visitor from another world notices the "pretty girl" rather as a Victorian bird watcher might observe courting birds. The only way to explain this behavior is to see it as competition for a female, and a female who, given the excess of suitors, has the power to exercise choice. The female chooses among differences, the differences accumulate (as divergence of character increases in the workings of natural selection), and thus without choice there would be no dimorphism in birds. Victorian as it distinctly is, this guess is a remarkably good one—rather, *because* of its self-evident Victorianism. Like a good novelist, Darwin here makes something rich and creative of his ideological luggage by way of an imaginative leap. He thinks himself into the bird's being.

It is worth pausing to notice, too, that in slipping into his

thought experiment and creating a hypothetical encounter between a pretty country girl and competing male suitors, Darwin
actually introduces into *human* mating the female choice that, for
humans, he was later to deny. The imagination in the form of a
thought experiment confirms the overall theory. But at the same
time it reaffirms the work of female choice, even among humans, although Darwin himself inverts the pattern for the human because he could not believe that human females had a significant role to play in the development of the species.

Darwin defines the excess in nature that cannot be explained
by natural selection as the "pretty" or the beautiful in the eyes of
the opposite sex. This is how to make sense of the astonishing
adornments of peacocks and pheasants, for example. The argument builds on his recognition that there are "pretty" things out
there (and we know they are pretty because we as humans regard them as so). If human beings find them beautiful (one piece
of evidence he gives is that women adorn themselves with bird
feathers!), then birds must too. Victorian birds.

Darwin's standards of prettiness in the *Descent* are un-
abashedly human, though he often concedes that what is beautiful to certain animals is not beautiful to humans. For the most
part, though, the evidence is based on human values, as he understands them. In his section on bird display, he invokes artists
to provide testimony to the amazing beauty of bird feathers and
courting habits. After describing the designs on the tail of an Argus pheasant, for example, he notes, "These feathers have been
shewn to several artists, and all have expressed their admiration
at the perfect shading" (2:91). In making his case for female
choice, he writes:

Many will declare that it is utterly incredible that a female bird should be
able to appreciate fine shading and exquisite patterns. It is undoubtedly a
marvellous fact that she should possess this almost human degree of taste,
though perhaps she admires the general effect rather than each separate detail. He who thinks that he can safely gauge the discrimination and taste of
the lower animals, may deny that the female Argus pheasant can appreciate
such refined beauty; but he will then be compelled to admit that the extraor-

dinary attitudes assumed by the male during the act of courtship, by which the wonderful beauty of his plumage is fully displayed, are purposeless; and this is a conclusion which I for one will never admit. (2:93)

Here again anthropomorphism is at work with a vengeance, but of course anthropomorphism is not usually for Darwin a mere sentimental lapse. It is rather a quite seriously worked out way of regarding a world in which there is an absolute continuity between humans and other animals. It is not so much anthropomorphism, then, as zoomorphism: that is, humans are animals, and therefore one can—as an animal oneself—understand non-human behavior simply by imagining one's way into the animal's mind.

This passage is only one of many that register Darwin's remarkable power to break through provincial prejudices by thinking through them. He can do this in part because he is entirely convinced, beyond theory and into the depths of his own imagination, that human feeling and thought are grounded in animal consciousness. Even worm consciousness, as we have seen. The cultural prejudices of the Victorians may be the cultural prejudices of worms as well, but engaging those prejudices and thinking with and through them allowed Darwin to think beyond them, as he did in the matter of female choice.

But there is a striking aspect to this argumentation that places Darwin even deeper inside the ideology of his own moment at the same time as it opens up new possibilities. To me, the most remarkable feature of this remarkable passage is that it participates in the same rhetorical methods as did the natural theology that Darwin had spent the best part of his research trying to dismiss. Dov Ospovat, some years ago, showed that the development of Darwin's theory depended in great measure on the influence of the natural theology that he ultimately rejected. Only after reading Malthus did Darwin come to the view (and this, too, gradually) that there was no such thing as "perfect" adaptation, and that the world, instead of being harmonious everywhere, was full of "discord"[16] But through this entire movement, the fact of "adaptation" continued to play a major role, and adaptation carried with it the old rational force of natural theology. Female intention, it seems, only supplements a nature already latent with the forms of a world that seems entirely intentional, infused, as Robert Richards has argued, with "deep Romantic strains" (*Romantic Conception*, 552).

At another point elsewhere in the *Descent*, Darwin says about a courting display, "We cannot believe such display is useless."[17] And in the passage I have just quoted he doggedly asserts that he will never admit that the courtship habits and style of the Argus pheasant are useless. Or consider this passage about monkeys: "It is scarcely conceivable that these crests of hair and the strongly-contrasted colours of the fur and skin can be the result of mere variability without the aid of selection; and it is inconceivable that they can be of any ordinary use to these animals. If so, they have probably been gained through sexual selection, though transmitted equally, or almost equally, to both sexes" (2:308).

For a convinced Darwinian like me, such argument, entirely representative of the passages on sexual selection, takes the breath away. In his *Natural Theology*, when he confronted the extraordinary contrivances of nature, Paley expressed incredulity that anyone could see these things and not recognize in them design and, necessarily then, the evidence of a Designer. Paley's designer is, of course, God. In discussing sexual selection, Darwin expresses precisely the same consternation and disbelief that anyone could deny the intention at work in the participating animals. The difference is that Darwin's designer is not God but female animals. Darwin argues this way despite the fact that the *Origin* is largely given over to demonstrating the error of imputing intention or design to an automatic natural process. The deep Victorian need for order and meaning, which found expression in late-century attacks on Darwin's theory of natural selection, is alive and well and useful in *The Descent of Man*. Whether in conscious imitation of Paley's methods or not, Darwin employs here the strategy of natural theology.

The question, "what is such and such a contrivance for?" is inescapable in evolutionary biology. In current controversy there is often debate about whether the ultimate benefit is the gene's, the organism's, or the group's (species'). But whichever answer one gives, the answers themselves imply an objective, a telos, or at least a quasi-telos. The contrivance may be "for" any one of these units, but it seems to be "for" *something,* and it is taken as worthwhile, or, rather, necessary. There may be no adequate answer; it may be, as in many instances Darwin described, simply the result of a correlation of growth of another adaptive part of the organism, a "spandrel," as Stephen Jay Gould called it. But whatever position one takes, the preliminary, thoroughly commonsensical, and therefore culturally loaded response is necessary. Something strikes us. We ask questions about it. It can only strike us if it fits into (or challenges) our culturally developed assumptions. Through sexual selection, Darwin introduces aesthetic taste and female intention as driving forces in the evolutionary process, and he could do that only because he had the cultural attitudes and assumptions I have been describing here.

Of course, this leaves out a great deal of the story. I have wanted for the most part simply to point to some things that look like paradoxes in the story of sexual selection and in Darwin's relation to that story in order to show that Darwin's theory is saturated with cultural values and feelings, but that he is enabled by the culture that is often taken to have led him astray. The exposure of that enabling, many good critics have claimed, demonstrates Darwin's ideological complicity with Victorian sexism and undercuts his "scientific" arguments, but such a reading assumes the dichotomy between "nature" and "nurture" that Darwin does so much to disrupt.

Exposing the complicity certainly tells less than half the story. Some of the rest of that story would reveal that those very ideological assumptions are not merely wicked transmitters of vicious ideologies but a condition of the really valuable and original work Darwin did. If as cultural critics and historians, we agree that there is no way to keep our present culture out of our most objective and serious intellectual work, and if we believe—as I think we must—that there is some intellectual work that genuinely changes things, then we can take Darwin's case as an important example of how conventional cultural views can, for a keen, observant, and thoughtful mind, open ways into new kinds of thinking, thinking that will disrupt the very conventions within which we are historically constrained to think. Darwin's theories, however they are locally hooked into what many of us might regard as ideologically repugnant aspects of Victorian culture, have in their afterlife no necessary connection to those nor to any other specific ideological positions. When other Victorian writers, particularly novelists like Hardy or George Eliot, use or play variations on Darwin's ideas about sexual selection, they arrive with their own ideological baggage and they produce other interpretations that might well generate other ideological positions.

Sexual selection is an amazingly inventive and productive idea. Following Darwin's mode of argument, it is hard to think about (or feel) the world as bereft of feeling and value. Even if it is sexuality that does it, the biological world is invested with the beautiful and works by means of choice. The act of imaginative sympathy by which Darwin manages to construct his theory of sexual selection is itself thrilling, both historically, as it runs against the culture's deep hostility to the idea of female choice, and aesthetically, as it finds precisely in the beautiful an explanation for the way things are. Intention comes back into the world, even if later thinkers try to lay over it the mindless model of an algorithm and it is intention driven by a strong feeling for the "pretty."

Deriving from a very Victorian notion of what is "pretty" and from a very Victorian sense of what is striking, Darwin's theory produces a scientifically interesting and likely correct understanding of evolutionary development. On the strength of it, one might make a case for the sexist Darwin as a kind of ideological hero in spite of himself. Certainly he believed in the intellectual inferiority of women. But female choice was about as revolutionary a concept as natural selection, and only recently has resistance to it diminished. That the theory of sexual selection is a product of Victorian culture is both unsurprising and remarkable since, if followed out to its fullest possibilities in directions Darwin established but did not follow, it might very well imply the intellectual superiority of women.

CHAPTER 7

A Kinder, Gentler, Darwin

Darwinism is blamed for taking meaning from the world by making
divine purpose optional. But Darwinism in much of its practice is a
project to populate the world with meaning, by identifying it in as
many aspects of life as possible.
—*Marak Kahn, **A Reason for Everything***

Despite Darwin's gentleness and compassion (and middle-class
gentility), despite his deep affection for his family and his kindness, despite the fertility of his imagination and the romantic
roots of his science, it would be, minimally, disingenuous not to
recognize the corrosive force of his thought, its power to drain
meaning from the world, its affinities with dog-eat-dog capitalism, and its uses in encouraging scientific racism and eugenics.
Even on the issue with which I have been directly concerned, the
question of "disenchantment," it would be absurd to insist that
Darwin's chance-ridden, mindless and heartless universe can be
felt to be as inspiriting as a divinely meaningful world, whose
worst elements might be reabsorbed into a theodicy based on
the idea of the fall. I have tried to attend to these aspects of his
work and life in earlier chapters. "There is no denying," says Dennett, "that Darwin's idea is a universal solvent, capable of cutting
right to the heart of everything in sight. The question is, what
does it leave behind?" (521)

My answer is, a lot. For Dennett, too, in a bravely Victorian
way, "some of these are losses to be regretted, but good riddance
to the rest of them. What remains is more than enough to build
on" (521). What will be built, on Dennett's account, will be ideas
that approach ever nearer, perhaps asymptotically, to the truth,
as the false faiths of the past are blown away by good science.
But Dennett's Darwinian passion for reason misses an important
element in Darwin, the quality of affect and of awe, the very

quality that William James found lacking in the Victorian positivists. Weber found it lacking too, and as a consequence developed his theory of disenchantment. In the long run the Victorians' experiment in radical scientific secularism didn't quite work, despite their ostensible passion for Dennett-like truth. As I have already noted, Positivism tried it and gathered very few aspiring souls. The freeman's worship, as Bertrand Russell called it, worked only for those rational secularists who worshiped, if anything, truth rationally derived. It may in the end be all that secularist naturalism leaves, all that the corrosive work of Darwin's theory allows to survive. But throughout this book I have been trying to suggest that it is not only Darwin's theory, in its abstracted form, that works, but the larger, connotative significances the language of its formulations bears.

In the preface to the second edition of *Darwin's Plots,* Gillian Beer, explaining why she had focused so intensely on the rhetoric of Darwin's writing, points out that some of her early readers were "puzzled," thinking that her concern with rhetoric implied that "Darwin's work is 'fiction.'" Her point was, rather, "that how Darwin said things was a crucial part of his struggle to think things, not a layer that can be skimmed off without loss" (xxv). Dennett, with whose interpretations of Darwin I am largely in agreement, is nevertheless a skimmer. Focusing quite reasonably on the direct meanings of Darwin's argument, he has little time to attend to the nature of the language in which it is couched.

For Beer and Robert Richards it matters a great deal that Darwin emerges from his texts as a child of romanticism. Beer calls him a romantic materialist. Richards argues that "from the very beginning, Darwin had recognized in nature a source of moral and aesthetic value" (*Romantic Conception,* 551). What "remains," when one has absorbed Richards' historical and rhetorical placing of Darwin's language, and after Darwin's arguments have done their corrosive work, is a world that I would want to call enchanted, alive with romantic spirit. Carlyle didn't think so, of course, and, before Darwin, complained through his Teufelsdröckh about the way the new sciences have turned the world into a mere mechanism, dead matter in motion. But, Richards says,

> the belief that nature was nothing but a vast machine and natural selection operated according to principles of a Manchester spinning loom—all of this remained quite distant to the mind that originally composed the *Origin of Species*. Darwin never referred to or conceived natural selection as operating in mechanical fashion, and the nature to which selection gave rise was perceived in its parts and in the whole as a teleologically self-organizing structure (534).

Look at the language, as Richards and Beer do, and another Darwin emerges. Richards attends to that language, particularly pre-*Origin* (but it's there in the *Origin*, too). So, without "skimming," I want to read the Darwin who, in his books and notebooks and letters, not only does not empty the world of value but who senses it as full of meaning, and who has access to it only because it is for him so richly laden with endlessly beautiful and endlessly various variations, adaptations, anomalies, and behaviors. I am proposing, then, a Darwin whose work at least partially justifies that bumper sticker that my son gave me some weeks ago: "Darwin loves you."

Putting aside questions of dogma, which belong to a different sort of book, it is safe to say that much religious resistance to Darwin derives from a sense that the world he describes is heartlessly and mindlessly bereft of a creator, an intelligent designer, and is thus meaningless, providing no consolations for its ravages. It is, rather, like the cruel "President of the Immortals" who presided, in Thomas Hardy's narrative, over the unjust death of Tess of the D'Urbervilles. In this chapter, then, I take up directly the argument for "enchantment" that was begun in the first chapter.

Among all the many different interpretations and uses of Darwin that one may find, there has been inadequate attention given to the Romantic Darwin whom Richards has most fully represented, to the Darwin whose language, as Beer has shown, is rich with affect and significances that might take us beyond Hardyesque indifference and cruelty, or the biological determinism that seems to have become central to so many contemporary interpretations of Darwin.[1] Darwin's science was not at all "disenchanting" or anesthetic or dehumanizing or amoral; it was rather a science enchanted from its inception with awareness of and awe at the complexities, varieties, beauties, and dangers of nature, and that was made possible by a deep romantic feeling for nature and its organisms.

Darwin's was a science that was only possible through an enormous effort of the sympathetic imagination, a science that is entirely compatible with precisely the sorts of feeling that many, ignorantly hostile to his arguments, believe can only be found in religion. If Darwin is to be seen as an apostle of secularism, someone who sought to explain the natural world entirely in terms of nature itself rather than in terms of the transcendent—certainly a legitimate understanding of him—, I want to represent him here as an exemplary figure in helping us toward a humane and sensitive secularity, one who refuses to minimize the cruelty of nature, but one who never loses a sense of its wonder, who never ceases finding objects of awe and natural reverence amidst its workings, and one who—to recall Jane Bennett's arguments about the possibilities of an entirely naturalistic orientation—continued to find in the midst of pain and loss, if not an "enchanted way of life," certainly "moments of enchantment" (10).

To be sure, the world in Darwin's hands is never animated by an intelligent designer. It remains subject to "laws" of nature rather than to some caring spirit, but those "laws" contain within them much that feels like "caring," and they are modified by their engagement with culture. The laws operate so as to produce a material reality that interacts with all those elements that have been taken as distinctively human: the mind, the spirit, the "heart." It is a world not of Teufelsdröckh's dead matter but of earth and organisms overwhelmingly beautiful, and difficult—terrible, lovable, and alive.

When Darwinism was saved by the "new synthesis" in the first third of the nineteenth century and Darwin was thus redeemed as scientist and world-historical figure, it became very tempting to disentangle his arguments from aspects of evolutionary thought that are incompatible with the best of current biology, or that might be regarded as ideologically suspect—those linking Darwin, for example, to modern forms of racism and eugenics. As Richards puts it, "to have [Darwin's] blessing on scientific positions one wishes to maintain in the late twentieth century can only advance their cause" (*Meaning of Evolution*, 176). I have been much tempted in my own great admiration for Darwin and his writing to exempt him from such things as "social Darwinism," and to assume—no scientist I—that he simply got it right on virtually everything for which, in his own time and place, he had the materials. But beyond that, he was so good an "observer" (as even he in the modesty of his autobiography admitted) that he was capable of making imaginative leaps that seemed beyond what the immediate evidence would allow. This was partly a consequence of his feeling for organisms, partly of the doggedness and precision of his observation, partly of his imaginative powers. How, I have wondered, could he have been so right about inheritance without knowing anything about genes or the Mendelian notion of particulate inheritance?

But Richards distrusts such celebratory reading of Darwin's powers, particularly efforts to read him into the best insights of contemporary evolutionary biology. He sees such interpretations as historically misleading. Neo-Darwinians, he says, "seem to have reached general agreement that three older proposals should be dismissed: that species evolution should be modeled on individual evolution, that embryogenesis recapitulates phylogenesis, and that evolution is progressive" (179–80). At the moment, denying that Darwin held these views has scientific and heuristic value, but Richards argues that Darwin's texts contradict the current orthodox opinion.

But I don't want here to argue for a Darwin who got it all right (as we understand the right) in his moment. My object, rather, is to try to recognize the ways in which his romantic imagination spurred his thought and informed it and gave to nature a significance that skimming misses. Nor do I want to argue through the complex historical/scientific questions that concern Richards, about the degree to which Darwin was in fact fully part of the tradition of teleological evolutionary thought. It is an interesting and difficult question, and I confess that my instinct has always been to follow the judgments expressed forcefully by Stephen Jay Gould, Ernst Mayr, and Peter Bowler. Richards rejects these largely dominant views by way of a close reading of Darwin in light of the tradition forwarded by Haeckel, that ontogeny recapitulates phylogeny.[2] Richards's understanding of Darwin's connection with this tradition allows him to foreground the romantic roots of Darwin's scientific practice and theorizing, and thus to hit upon the element of Darwin's work that I, with a strong ideological inclination of my own, am trying to argue for.

I am not making a case here for Darwin as a teleological and progressivist thinker (although the evidence Richards brings forward for this still largely unorthodox view of Darwin is strong). But Richards is surely right about Darwin's romantic roots, and recognizing these should provide considerable justification for attempting to wrest Darwin away from those who say that secularism and naturalism are arid, spiritually barren ways to approach the world. Darwin does not imagine the world as a mechanism, nor does his rigorous science dispel spirit and value from the natural world. It may be a leap from this incontrovertible fact to my effort to represent Darwin as a potential model for a thoroughly, radically secular, but affectively and aesthetically and morally satisfying, understanding of the world, but that is the fundamental effort of this book. As opposed to the inadequate alternatives—a religious view of the world that offers to understand and explain it primarily in transcendental terms or a scientific view that reduces biology to mere mechanism—I propose here a Darwin who, while absolutely and pervasively materialistic in his view of how the world works and of how humans got to be what they now are, does not reduce the material to a mere mechanism from which value and affect are entirely expelled by formulaic processes.

I agree with Richards that it is virtually impossible to take the ideology out of historical representations, but that one can come more or less close to an adequate, a "Rankean," representation of what Darwin was really saying and in what intellectual context he was saying it.[3] On the one hand, then, there is the Darwin who fits neatly into the Enlightenment project, who probably more than any other writer or thinker enabled the scientific study of the human that had been the primary objective of nineteenth-century positivism. This positivist Darwin brought to the world the undeniable news that we are, as Matthew Arnold satirically quoted him, "descended from a hairy quadruped, furnished with a tail and pointed ears, probably arboreal in its habits." (*Descent*, 2:389). For this Darwin, all organic change can be explained by sequences of small natural causes, and even human behavior has its roots in the heartless processes of evolutionary development.

On the other hand, there is the Darwin we have met in the first chapter, who set out on the *Beagle* voyage in part because of his romantic fascination with the wonders of tropical nature, and who on the very first page of his *Beagle* diary, describes the extraordinary happiness he felt going ashore during the ship's first stop at Porto Praya. This is not the weary, disenchanted, long-bearded Darwin of the most famous images but an exuberant young lover of nature, feeling the enchantment he has long been accused of banishing. *That* Darwin survived to old age, delighting in the seeds, ants, worms, and miracles of his own garden.

This happier, kinder, gentler Darwin is a figure widely recognized: virtually all biographies regard him as a modest, gentle, and loving man, and, though the quality of his character is not my major point, it remains relevant.[4] This niceness has sometimes been taken as a symptom of his unself-conscious bourgeois sensibility and has evoked the critiques that reasonably enough point to the kinship between his theory and laissez-faire capitalism, and that emphasize the sexism of parts of *The Descent of Man*, as we have seen in the preceding chapter. But the

Romantic Darwin was, indeed, a man of his moment, and recognizing the man in the theory is an important aspect of perceiving the kinder, gentler Darwin I am trying to evoke here.

My main point, however, will be that his writing itself, not simply in its literal meanings (no matter how complex they might become), but in its very texture and writerly quality, projects a relationship to the natural world that is full, dare I say it, of reverence and awe. It offers an unblinkered (and only occasionally sentimental) view of nature, a view that, if widely shared, would, I believe, radically change for the better the conditions both of nature and of humanity everywhere. Nobody could reasonably argue that we should all now believe everything Darwin said; it is important, however, to attend to the way he said it. Darwin has been put to terrible uses, but the nonsecular alternatives to his vision, ranging from Christian fundamentalism to kooky spiritualism, obviously won't do. But Darwin himself, read differently, might well be the best antidote to the uses of Darwin in the great rationalizing and dominating tradition of Western science.

The literature of science is not the most obvious place one would go to look for romantic inspiration, and Darwin certainly struggled through virtually his entire career to make sure that his writing was factual, generalizable, and scientific within the terms that his own generation would value. His writing is often awkwardly impersonal in its aspirations and seems sometimes to do somersaults to say in the passive voice what would come much more effectively in the active. He wants, very clearly, to leave himself out of it, but only occasionally succeeds. (That strategy itself, as students of nineteenth-century objectivity have demonstrated, is part of a widespread and highly moralized tradition of selflessness, so that to be scientifically accurate is to practice the highest morality.) Yet virtually every reader of his most famous works is struck by how much a personal voice is present and how frequently that voice slips into expressions of admiration and even awe—certainly not the sort of thing one would be likely to publish in a professional paper these days.

But it is not only the voice. It is the imagination and the sense of a world animate, in movement, and full of creatures profoundly akin to ourselves.

The two Darwins I have adumbrated require that I first take a look at his struggle for "objectivity," his attempt to make sure that we understand that his arguments are founded on as much evidence as possible and that the truth of these facts would immediately "strike" anyone who looked. This strategy is obviously akin to that of the modern scientific tradition of detachment and disinterest. The world is what it is, not an object half-perceived and half-created, bending to the observer's interests and desires. (We as readers may well feel that Darwin's world, so imaginatively conceived and partly on the basis of many thought experiments, is, after all, half-perceived and half-created, while we can continue to recognize that what is "created" by consciousness is close to the richest possible imagination of the nature of what is perceived.) In this, then, there is no escaping the question of "disinterest."[5]

But second, I will want to pick up again, after the discussion of sexual selection, what Evelyn Fox Keller has called "a feeling for the organism," the qualities of vision, imagination, and learning that Keller discusses as marking the life and work of Barbara McClintock. Keller quotes McClintock explaining what it was that allowed her to "see further and deeper into the mysteries of genetics than her colleagues." The answer: one must have "the patience to 'hear what the material has to say to you,' the openness to 'let it come to you.' Above all one must have 'a feeling for the organism.'"[6] Reading this about McClintock, I thought immediately of Darwin, one of whose supreme virtues was precisely his power to listen to nature in this way, to open himself to its differences and strangeness. Darwin, as we have already seen in the discussion of the Argus pheasant in the previous chapter, sought in an almost literal way to know what it was like to be the organisms he discussed, and—as, for example, his eight-year excursion into the nature of barnacles demonstrates— he had precisely the patience required, precisely the imagination to understand what those barnacles were up to, precisely the passion for the organism that allowed him to devote his life to them. As for the child of his brain, natural selection, for Darwin there was no detail of its operation too minute for his attention.

It is true that, like McClintock, Darwin was almost singular in his intense engagement with the natural world and his obvious excitement about it, and love for it. And while it is hard to imagine the lay reader engaging in the details that such patience and imaginative risk-taking dig up, it is nevertheless obvious that Darwin found the sort of spiritual sustenance he needed in the natural world itself, in the very miracle of its complexity and variousness, in the overwhelming beauty of its manifestations, and in the wonder of its precision: so bees, Darwin shows, "have practically solved a recondite problem, and have made their cells of the proper shape to hold the greatest possible amount of honey, with the least possible consumption of wax." As he enters upon the subject, Darwin exclaims: "He must be a dull man who can examine the exquisite structure of a comb, so beautifully adapted to its end, without enthusiastic admiration" (*Origin*, 224). It is with enthusiastic admiration, even in the passive voice, that Darwin imagines and describes the ostensibly heartless nature he observes. Darwin's world is not neutral, despite the absence from it of some transcendental being. It is laden with value in its smallest cell and its most minute larva.

I

To begin with, then, and partly to avoid an easy sentimentalizing of Darwin's relation to his subject and to science, I want to consider his aspirations to self-effacement. There is a curious parallel between the power of religion to transcend the self, to absorb the self into a larger entity, and the power of Darwin's science, at least theoretically, to absorb the self into the natural world. The affect is certainly radically different, and there are no obvious rituals that go with scientific selflessness. But it may be possible to conceive the enormous enchanting power even of science when we consider the demands it seemed to make on the observer to quiet personal desires, to diminish the appeals of the self. And yet in thinking about those demands, we may find it possible to understand better how valuable, one might even say "sacred," the natural world was for Darwin.

Very few people now believe the story of scientific disinterest, at least not with the intensity with which it was believed in the nineteenth century. Foundationalist positivism has, through the entire last half of the twentieth century, been badly discredited, and while science has proceeded much as it did before, in cultural critique and most historical study of science the whole tradition of a disinterested approach to things has been so thoroughly undermined that any apparent manifestation of selflessness tends to be treated with deep distrust and assaulted by historical and critical efforts to demonstrate the selfishness under the modesty. Darwinian theory itself, in the guise of sociobiology and evolutionary psychology, has made it extremely difficult to believe in authentically selfless action. "Reciprocal altruism" is often the best we can do. Darwin's modesty, which won him many friends in his own time, has gained for him much distrust in ours, and the modesty has been connected with the traditions of disinterest and objectivity that were particularly strong in nineteenth-century science, and in a way constituted its official self-description in its developing professionalization. Darwin's *Autobiography* dramatically embodies this attitude in the modesty of its self-representation, and Darwin's theory certainly does emphasize processes that progress impersonally. It does not require profound knowledge of Darwin's theory to recognize, for the most obvious example, that natural selection partakes of many of the qualities characteristic of what was, in the nineteenth century, orthodox Baconian empiricism (a mode of perception from which all the "idols" have been banished).[7] The good empiricist, it should be remembered, is one who labors, as one of Darwin's favorite books argued, to overcome, in his registering of "experience," all prejudices of sense and prejudices of opinion. The prejudices of sense are *inevitable* but can be counterbalanced, if only by rigorously disciplined efforts to attend impartially to experience (Herschel, 67–84).

Nineteenth-century observers often intuited something like this about Darwin's work. In his notorious "Belfast Address," John Tyndall, for example, discusses Darwin as a model scientist. One of the conditions for Darwin's success, Tyndall argues, is his exclusion of personal emotion: "Mr. Darwin shirks no difficulties" in his determination to get it right at any cost, and he seeks no mere dialectical victory. Rather, he wants "the establishment of a truth which he means to be everlasting." The most sustained and irritating attacks do not irritate him:

> He treats every objection with a soberness and thoroughness which even Bishop Butler might be proud to imitate, surrounding each fact with its appropriate detail, placing it in its proper relations, and usually giving it a significance which, as long as it was kept isolated, failed to appear. This is done without a trace of ill-temper. He moves over the subject with the passionless strength of a glacier; and the grinding of rocks is not always without a counterpart in the logic of pulverization of the objector. But though in handling this mighty theme all passion has been stilled, there is an emotion of the intellect, incident to the discernment of new truth, which often colours and warms the pages of Mr. Darwin.[8]

Tyndall is a publicist as well as a scientist. He creates Darwin here as the ideal scientist by not attending to the various ways in which his "grinding" is interrupted by exclamations of pleasure and excitement, and by admissions that he has not enough evidence, or that an alternative view would be fatal to his theory. But clearly Tyndall is aware of that fundamental quality of virtually all of Darwin's work: the excitement with which he pursues truth "colours and warms the pages."

Tyndall's exaggeration of Darwin's self-suppression is less over the top than Darwin's own, in which the notion of "grinding" is also, famously, invoked, although with emotional force and regret. "My mind," Darwin notoriously lamented, "seems to have become a kind of machine for grinding general laws out of large collections of facts" (*Autobiography*, 138). This seems a long way from his romantic youth, but much in the *Autobiography*, even its determination simply to lay out a series of anecdotes and facts for the possible interest of his family, belies the self-description, not only at the point at which Darwin confesses to liking novels with pretty heroines and happy endings, but in the famous opening, in which he claims to be writing as if he were "a dead man in another world looking back at my own life" (21).

One should not underemphasize, however, the professional usefulness of Darwin's characteristically selfless mode of dealing with the world. The "selflessness" was surely part of the reason for his attractiveness to others. He wrote a letter to William, when the boy first went off to school, suggesting that "you will find that the greatest pleasure in life is in being beloved; & this depends almost more on pleasant manners, than on being kind with grave and gruff manners. Depend upon it, that the only way to acquire pleasant manners is to try to please *everybody* you come near, your school-fellows, servants, and everyone" (*Correspondence*, 5:63).

Lovableness and science came together in Darwin's life: through self-deprecation and gracious strategies of generosity, he managed to encourage a vast network of scientists and naturalists throughout the world to think of him and work hard for him. He was not at all above thinking of the scientific advantage he might gain from leaning on others to provide evidence, to invoke his work, to send him specimens. Sorry as he expresses himself to be about bothering people for information, he never hesitates to do so. Darwin's selflessness was also a condition for the personal authority he developed and for his successful scientific practice.[9]

It would be to overestimate, however, the degree to which he instinctively minimized himself both in his own mind and in relation to others whose love he wished to evoke. The concluding sentence of his autobiography suggests something of this: "With such moderate abilities as I possess, it is truly surprising that thus I should have influenced to a considerable extent the beliefs of scientific men on some important points" (145). In Donald

Fleming's suggestive discussion of Darwin's crisis of feeling, as he described it, he calls Darwin an "anaesthetic man."[10] The anaesthesia was for Darwin a moral, aesthetic, and political condition, and his analysis that this resulted from his scientific "grinding" must have some serious basis. Although, as I have suggested, Darwin was far from "dead," the entire process by which he managed for years to submerge himself in his scientific work while at the same time he was losing his beloved daughter and observing, rather silently and most often from a sick bed, the battles that Huxley and his friends fought in his behalf—all of this does seem to have had some effect in deadening his sensitivity to literature, to the Wordsworth and Milton whom he had loved, to Shakespeare, to poetry almost entirely. The passion that Darwin had originally felt for the world that Humboldt, for example, had opened to him, seems to have been sustained but muted by the fact that he was, from the time of the *Beagle* on, virtually confined to his house and garden. The sublime of the tropics became the excitement of specimens. He acquired the habit of looking with steady detachment and with as much distance as possible at the details of nature's processes through the specimens that others were sending him. The transformation from a vital, active, daring, and energetic young man into a valetudinarian at the age of thirty is paralleled by the development of his thought. Beginning with wonder at the marvels of nature, Darwin goes on to take his greatest pleasures from his attempts to explain them.

Along the way, he continues to apologize for allusions to himself. Note a letter to Adolf von Morlot: "with respect to passages in my letters, I feel almost certain that they contain nothing new or worth publishing; and they were written without care and with personal allusion to myself, which are not fit for any eye, but a correspondent" (*Correspondence*, 3:88). Darwin had his strong personal opinions and feelings and would occasionally express them in letters to friends and notes to himself. But in his public writing, as I have already remarked, he struggled hard to keep himself out of the text. The pleasures expressed more or less openly in the diary for the *Beagle* voyage tend to disappear.

And yet, however distanced Darwin tried to become from himself, it is hard to read him without being aware that the "anaesthetic man" was quietly aglow with an aesthetic appreciation of nature. It was nature that, finally, replaced Shakespeare and the poets in his aesthetic responses. "This is a marvelous world we live in," he wrote to a friend, "and I never cease marveling at it" (*Correspondence*, 2:125). If anything is permanent in Darwin's evolving world, it is this sense of the marvelous: it recurs in the last paragraph of his last book, in which he reflects with wonder on the work that worms do.

Darwin's fascination with minutiae, his tendency to think in gradualist terms of the world transforming as the result of slow and minute changes, is always attached to an emotion of wonder. Darwin's sublime, as I have noted elsewhere, emerges from his grinding, his tedious, dispassionate examination of the ostensibly trivial, and his shocked recognition of what the accumulation of minutiae produces. Wonder was the beginning and the end of Darwin's work, in which, it almost seems, he is taken by surprise by what his careful, cautious, disinterested investigations reveal to him.

II

James Paradis has valuably described Darwin's movement from "the aesthetic idealism of Romantic art" to "the system building traditions of geological and natural sciences" in the writings about the *Beagle* voyage, and he argues that "Darwin traced the historical path from poetic to scientific nature in his M transmutation Notebook in 1838" (85, 100). An essentially poetic response to the natural generates a scientific one—first poetry, then science (though of course in the young Darwin the two were completely contemporary), but in the latter phase, as this essay is concerned to suggest, poetry and science existed not independently but in a new kind of balance. The work on the *Beagle* materials does suggest a concentrated effort to move beyond the personal intensity of Darwin's first engagement with the tropics and other parts of the world toward the formulation of "general laws."

The notebooks are full of entries that are clearly aspects of the aspiration beyond wonder, the attempt to transform the ostensibly miraculous into the ordinary. As a practical consequence of his commitment to Lyellian uniformitarianism and actualism, this works as a strategy to overcome the arguments of natural theology: everything can be explained by causes now in operation. We need no epochal catastrophes, not even an intervention from God. The miracle, then, rests precisely in the fact that the wonders of the world are all the product of the ordinary. No vision is more fundamentally romantic than this one. Wordsworth remains alive and well in the very texture of Darwin's gradualist materialism.

Rebecca Stott captures something of this quality in describing a passage (part of which I will quote below) from his work on barnacles. She points out how Darwin "found it impossible to maintain the critical distance needed to weigh up facts rationally and posit hypotheses" when he tried to describe his discoveries about "the sexual peculiarities" of various barnacle families. The barnacle books are, as Stott points out, "the most considered and dry of the texts that he would ever write." But "wonder would overtake him" as he reflected on "the microscopic communities of creatures he found living within a single sac of a single barnacle" (141–42).[11] The passage Stott quotes exemplifies the qualities discussed here. It is, moreover, centrally Darwinian, both in its dramatization of the loss of self in the close inspection of nature and in the exuberance and excitement—manifested by the almost inevitable Darwinian exclamation point—with which the catalogue of observations is set down:

> As I am summing up the singularity of the phenomena here presented, I will allude to the marvelous assemblage of beings seen by me within the sac of an Ibla quadrivalvis,—namely, an old and young male, both minute, worm-like

destitute of a capitulum, with a great mouth, and rudimentary thorax and limbs, attached to each other and to the hermaphrodite, which latter is utterly different in appearance and structure; secondly, the four or five, free, boat-shaped larvae, with their curious prehensile antennae, two great compound eyes, no mouth, and six natatory legs; and lastly, several hundreds of the larvae in their first stage of development, globular, with horn-shaped projections on their carapaces, minute single eyes, filiformed antennae, pros.bosciformed mounts, and only three pairs of natatory legs, what diverse beings, with scarcely anything in common, and yet all belonging to the same species!

Here is a description to make Balfour squirm, and yet the intensity, precision, and excitement of the technical language testify to a scientist who finds in the world's least creatures grounds for wonder. We have again that reverse sublimity that is so central to the experience of reading Darwin. Inside the minute sac of a microscopic organism is "a marvelous assemblage of beings." And the only way to convey the marvel is to be as precise and particular, as scientific as possible. So while, on the one hand, Darwin doesn't hesitate to structure the whole cluster of details inside a frame of wonder, those details are piled on with a remarkable care that requires Darwin the writer to step back from himself and his own wonder. The wonder is conditional on the disappearance of the self.

So in his A Notebook, he makes a point that he will develop more fully in the *Origin*, that "There is no more wonder in extinction of species than of individual" (*Notebooks*, 63). A more complex entry in Notebook B is concerned with the means of transporting seeds or eggs over great distances. Darwin asks if there are "any genera, mundane, which cannot transport easily, it would have been wonderful if the two Rheas had existed in different Continents. In plants I believe not" (*Notebooks*, 196). The point here is that wonder depends on the humanly inexplicable happening, and the transportation of numerous rhea eggs across continents could not be explained as Darwin could explain the transportation of seeds. In both of these entries, the scientific issue is created in response to the capacity of wonder to intimate a nonsecular mechanism. Darwin's response is to find ways to intimate not the transcendent but what Teufelsdröckh would call the descendental—tiny and not very beautiful organisms, which can be described with no romantic vocabulary at all, but which, in their very apparent insignificance, evoke wonder also. If we take the death of the individual as "natural," so we must take the death of species. And there is nothing wonderful about the fact that rheas do not exist on two different continents. Similarly, there is nothing wonderful about the fact that the same species of plant do exist on two different continents—once, of course, one has figured out through experiment the various ways in which seeds might be transported.

Darwin's arguments for the implication of humanity in his evolutionary scheme depend, as well, upon his capacity to diminish the official wonder at man's moral and intellectual nature. That is, he needs to show that all of the qualities taken as peculiar in the human species are in effect present in less exalted species in the natural world. The M Notebook, which Paradis takes as suggesting the fullest transition from poetry to systematic and impersonal science, are full of entries that bring the human and the nonhuman into contiguity. There is even an entry on the free will of dogs, and if dogs "then an oyster & polype (& a plant in some senses, perhaps, though from not having pain or pleasure actions unavoidable & only to be changed by habits)." The discussion of "free will of oyster" achieves something of the comic sublime, but it leads Darwin to consider the degree to which free will and change are related to biological "organization" (*Notebooks*, 536).

The strategy of diminution in Darwin is ultimately a strategy of sublime displacement of a more traditional sublime. Everywhere in his work, almost from the start, he attempts to find explanations for remarkable phenomena that naturalize them, put them within the comprehension of humans. Ironically, however, and this is one of the keys to the aesthetic and moral complexity of Darwin's work, the very possibility of such reduction is sublime—that is, it shocks the common intuition that large effects much have large causes, that sublime feelings must derive from the transcendentally spiritual. All of this is of a piece with Darwin's own strategies of self-effacement. Darwin resists the sublime and creates it at the same time, and he depends in part on the shock of discovery, that the tallest cliffs are the product of millennia of slow incremental rises, that the life of isolated islands begins not with a "fiat" but with grubby little insects, that the heights of human moral sensibility derive from the sexual activity we can recognize in peacocks and peahens.

For Darwin, phenomena must be reduced, as within the empiricist doxology, to "sensible" ones, although what is required is ultimately a particularly expert "sensibility," one which can see behind the experience of, say, the same plant on two continents, into deep historical time and understand how seeds might be transported over vast tracts of ocean, or how the sensory equipment of an oyster is cousin to our own. Anything transcendent of experience is at least ostensibly excluded. But the power of ordinariness to produce what we have often taken as transcendent achieves a kind of sublimity in itself.

Darwin's scientific self-effacement is not a disguise for unacknowledged feeling and ideology but part of a complex romantic strategy that sees self-abnegation as a supreme moral virtue allowing access to the great world of the not-self. Darwin's lifelong passion for science was driven by deep personal and romantic desires, and issued out into them, though in another form. From sublime to sublime: and along the way the self must diminish itself and be diminished by the objects it perceives. But the Darwin who was nauseated by Shakespeare did not see the world as a mere impersonal mechanism, or regard it, from his detached standpoint, as a brutally inhumane process that was regardless of the fate of the human. It is not beside the point to say that while most of his culture was appalled at the idea that humans were related to the other primates, Darwin, out of a deep humility that was part of his passion for the natural world, was not bothered at all. The fact of the emergence of human intelligence from sensibilities like that of the oyster was for him a miracle of another sort, one built out of natural law. After all, as

Whitman was saying across the ocean: "A mouse is miracle enough to stagger sextillions of infidels." The infidel Darwin felt the power of that mouse, and for him it was a condition of that power that he remain infidel.

I have been emphasizing Darwin's Romanticism as a kind of antidote to the dominant cultural view that he is one of the prime culprits in the modern disenchantment of the world. In a superb brief chapter about Darwin, Richards connects him with German romanticism, with Humboldt, whose writings are a clear inspiration for the *Voyage of the Beagle*, and with *Naturphilosophie*, which regarded nature as teleologically structured. Expelling the god of British natural theology from nature, Darwin, according to Richards, built a model of nature like that of *Naturphilosophie*, in which the creative powers of God are shifted to nature itself, whose "creative power," as Richards discusses it, works "in the gradual, evolutionary unfolding of telic purpose" (518). Richards knows quite well that much of Darwin's argument depends on dispelling from the evolutionary process any "telos," and yet current scientific emphasis on the rejection of teleology is part of what Richards regards as an ideologically driven distortion of Darwin. But wherever Darwin's theory may have eventually taken him or his followers, Richards insists that his ideas flourished within or on top of a system that was initially teleological, and that one can see reflections of this even in his final formulations. Telos or not, it is clear that Darwin's nature is nothing like the mechanical model of it we are used to hearing about. "No phrase," says Richards, "comes so trippingly to the lips of contemporary biologists as 'the mechanism of natural selection.'" But this mechanical view is not Darwin's.

Richards points out that the word "machine" appeared only once in the *Origin*, and it is clear from the nature of the metaphors that Darwin decoyed in constructing and arguing for his theory that he read nature as animated, not steam driven. The tree of life may seem a banality, but it is a critically precise analogy that Darwin uses at the end of the fourth chapter of the *Origin*. While the "mechanical" inferences from his theory by many post-Darwinian thinkers are entirely understandable, Darwin did not build his theory by emptying the world of creative energy. His world is, as Richards insists, vital and creative. The wonder that Darwin constantly felt is not at a spiritless automaton but at the astonishing power of nature to create out of its smallest entities complex and morally sensitive beings, and to fill it, as it fills the Ibla barnacle's sac, with wonderfully various and creative beings.

The tendency to "mechanize" Darwin is overt in Daniel Dennett's emphasis on the austere, corrosive logic of Darwin's rigorous arguments, into which every detail of *The Origin of Species* is made to fit. Dennett's reading, which we had occasion to discuss briefly elsewhere in this book, needs further attention. It is so strong and convincing a reading that it is bracing, as well, and in its translation of Darwin into the situations of modern polemic, it is largely accurate. But in making his high moral case—a case that, I have argued, echoes the Victorian "free thought" movement in the last third of the nineteenth century—against the intellectual and moral weakness of those who need "sky hooks" to help them to face the hard facts of natural world, Dennett is being true to one strain of post-Darwinian thought. I want, however, to consider here the language that makes his argument affectively untrue to Darwin, despite its accuracy and cleverness and severity.

Dennett's Darwin fits the kind of argument Carolyn Merchant has made, that modern Western science turns nature into a mechanism—the kind of mechanism against which Carlyle raged in Darwin's own time.[12] This way of viewing nature, she claims, always entails subjection and exploitation. Dennett's translation of Darwin—which is only, I think, a more radical form of a widespread way of understanding him—mechanizes nature's processes and describes nature as an object to be manipulated, dead. It is, then, just the reverse of the nature that Richards describes in his discussion of Darwin's romanticism.

Darwin's language gives us a nature active and entirely alive, and it marks him as a direct heir of English romanticism, participating in just the Carlylean world that distrusts the hard, rationalizing work of science and produced Blake's stunning, paradoxical image of a nude, muscular Newton bent over his compass, rounding off the world far too geometrically. Darwin too sees the world in a grain of sand. Or in the spots on a beetle's back, or in a barnacle's sac.

What can't be skimmed off is the metaphorical play of language that infuses nature with independent and organically intricate life, that analogically builds trees of connection, that reverently describes the miracle of diversity and development, and that ties living nature with earlier mythic worlds. Though undeceived by it, Darwin remained in love with nature, which drew on the deepest resources of his mind and of the language that from Milton through Wordsworth had endowed it with a mythic energy while registering it with startling precision.

Dennett is one of the most insistent of "skimmers," as he translates Darwin's argument into the mechanical metaphor of an "algorithm." The move would seem to be justified, following as it does Darwin's unequivocal refusal to find explanation outside of "secondary causes" (488). Darwin, however, gives us not "algorithm" but "natural selection," a very different, if obviously related, thing. Virtually every semipopular essay about Darwin ends with an allusion to the famous last lines of the *Origin:* "there is grandeur in this view of things." It may seem to many hard-nosed modern Darwinians a mere throwaway line after four hundred pages of relentless secularizing, but it is important, because it embodies that crucial aspect of his overall argument, an inescapable affect that goes with the counterintuitive and reverse sublime vision that I have been discussing.

The difference between Dennett's formulation and Darwin's, might be suggested by the way Dennett, on the one hand, makes "algorithm" do the work of eliminating all the metaphorical development of Darwin's argument, and, on the other, creates "algorithm" as a metaphor for mindless and mechanical activity. The mindlessness of natural processes, which Dennett wants to emphasize, is, of course, derived directly from Darwin's emptying of the universe of any transcendent planner, and it is central to the work of giving to the theory irrefutable authority, and withdrawing it absolutely from the realms of caprice, chance, doubt, and vague emotive vacillation that mark normal thinking. It might be seen as an equivalent of Darwin's continuing struggle toward selflessness, which, however, as I have tried to show, was always entangled in complicated personal commitments.

An algorithm, Dennett notes, entails *substrate neutrality* (that is, it works logically whatever the material incarnation), *underlying mindlessness* (that is, it is absolutely simple in its sequence of steps, none of which entail intelligent direction), and *guaranteed results* (that is, it work always, everywhere, unerringly—it is, in other words, "a foolproof recipe"), (51).

Useful as this is to evolutionary psychology, it is inadequate to Darwin, whose language—even as in revision he fought his own instincts—always turned natural selection into a caring, attentive, active personification. But it is important not to sentimentalize. There are intellectual grounds for accepting Dennett's reduction of Darwin's prose, and we should recognize too how easy it has been to fold Darwin into a narrative of the imperial vision of modern Western science, a narrative that Dennett himself would probably despise but that is surely the other face of the algorithmic Darwin—the face that reveals the ideological forces disguised by selflessness. Much has been made in recent years of the ways in which Darwin domesticated—or, perhaps, bourgeoisified—evolution, pulled it from the hands of the politically radical types who had inclined to support it, and made it reflect, as Engels said, the laissez-faire economics of Darwin's own class. The complicity of scientific investigation, particularly of natural history, in the work of empire, has been argued by Mary Louise Pratt, who talks of the Enlightenment interest in "systematizing nature" as a "European project" entailing the "circumnavigation" of the world and "the mapping of the worlds' coastlines" (30).[13] No accident, then, that Darwin's crucial voyage on the *Beagle* was explicitly part of that mapping enterprise, and it would be disingenuous to attempt to disconnect his natural history from the work of empire as Pratt describes it.

Surely, a science that scrupulously obscures its assertions of domination in the anonymity of disinterest and slides into algorithms identified precisely as being unmarked and impersonal is a perfect disguise for domination. But, as in considering Dennett's accurate but inadequate rendering of Darwin's prose and his way of thinking, I want to turn back to the kind of language Darwin used in making his case, and the way that language sometimes metaphorically, sometimes simply by virtue of its particularity and intensity of attention, struggles to avoid allowing Darwin to impose his own mind and desires on nature—precisely, I would argue, because he loved it so intensely. His passion is not to dominate nature, or to exploit it, but to listen to it, open himself up to its strangeness and difference, find ways to cope with its sublimity.

III

Darwin's attempts to let nature speak for itself, which suggests a kind of submersion of self, does not diminish one of the striking things about *The Origin of Species*: the sense it conveys of being argued in a personal voice, one easily moved to excitement and at ease in the messiness of homely details. No important scientific text is more thick with exclamation points than the *Origin*. The chapter called "Struggle for Existence" is sprinkled with exclamations of wonder, both at nature's overwhelming richness and at human error in understanding it. "We invoke cataclysms to desolate the world, or invent laws on the duration of the forms of life!" Darwin exclaims, when extinction can often be explained by simple, ordinary causes.

But beyond exclamation points there is Darwin's larger argument, as when he develops with great precision and particularity that metaphor of the tree to describe "affinities of all the beings of the same class." Tracing through the metaphor, the branching, layering, leafing of a giant tree, Darwin famously concludes: "As buds give rise by growth to fresh buds, and these, if vigorous, branch out and overtop on all sides many a feebler branch, so by generation I believe it has been with the great Tree of Life, which fills with its dead and broken branches the crust of the earth, and covers the surface with its ever branching and beautiful ramifications" (172). The Darwin who ended by being nauseated by poetry created his own on his way to his theory. The image of "branching and beautiful ramifications" has the etymological precision and alliteration and reverence for life that poetry might offer, right in the midst of a description of what is so often taken as the radically indifferent processes of nature.

Darwin's anthropomorphism, which we have encountered frequently earlier in this book, is another aspect of his enchanting prose. We have already seen how, in the development of his theory of sexual selection, anthropomorphism did valuable service, was not merely a layer "to be skimmed off" the theory. Anthropomorphism is no mere metaphorical flourish. It was partly because he had so intensely that feeling for the organism that he was so quick to project human qualities into everything from worms to dogs, and that he was able to imagine humanity as on a biological continuum with all other animals. Moreover, the anthropomorphism helped him argue for the biological continuum with no sense of loss but with excitement and pleasure. John Durant makes the point succinctly: "Darwin had no hesitation in using his own species as a source of insights into the rest of the world of life. In this sense, the idea that man was an animal, to be studied and known in the same way as any other, was not so much a conclusion of the theoretical endeavors of the notebooks as it was a precondition for them."[14]

Who else, then, would have thought of playing the piano for worms, or blowing smoke at them? Or listen to the language with which he describes one of his little domestic experiments: he talks of how he "prevented the attendance" of ants on aphides, then "tickled and stroked" the aphides "in the same manner . . .

as ants do with their antennae" and then "allowed an ant to visit them" (211). Darwin's most "disenchanting" move—the denial that humanity (or any other species) was a special creation of God—was largely enabled by an anthropomorphism that might equally well be considered an act of re-enchantment. Darwin's nature is populated by creatures whose behavior can be inferred through human projection; his anthropomorphism, as I have already shown in discussing his theory of sexual selection, becomes a remarkable act of the sympathetic imagination.

At the end of his life, Darwin was working on worms. His book on "vegetable mould" and worms, in which, it appears, he assumed nobody would be interested, sold faster than any of his others. It is a fitting culmination to his career that the reductionist method that he learned from Lyell at the start of his career should come to enchant so wide a reading public. And his good friend Hooker, as quoted by Janet Browne, excitedly wrote: "I must own I had always looked on worms as amongst the most helpless and unintelligent members of the creation; and am amazed to find that they have a domestic life and public duties!" (*Power of Place*, 490). They are, Hooker realized, like Darwin's bees, perfect Victorian citizens.

Darwin's science was intricately intertwined with his life and his love of poetry, but also with his scientific ambitions, and these might seem to have pulled him in different directions. While he saw animals and humans on a continuum, as a scientist he was strongly opposed to the antivivisection movement. On the one hand, then, Darwin's intense sympathetic imagination led him to feel for worms more than most people feel for whole forests full of animals. As Janet Browne argues, he was appalled at the exploitation of the weak, and he seems to have connected inhumanity to animals with slavery. It is this characteristic combination of attitudes conventionally taken to be mutually incompatible that marks Darwin's work and his life. On the other hand, "Disgusted by cruelty to animals of any kind," as Browne describes him, and thus severe, in his role as magistrate,

on anyone who mistreated animals, . . . he was also totally dedicated to what he thought of as 'the ideals of pure research'"
(420) and went on to resist forcefully the antivivisection bill designed to regulate physiological laboratories.

Darwin's works are not romantic rhapsodies, though they are
at times metaphorical and poetic. The project Dennett describes
is, one must concede, very much an aspect of his writing, but it is
worth pausing for a moment, to look not at passages in which he
expresses wonder at his subject, but at passages in which he is
simply doing his work in a language that aspires to pure literal-
ism and to a detachment that even excludes his frequent exclamations of wonder. And there is no book more ostensibly deflat-
ing of human dignity and spiritual significance, more "objective"
in its self-representation, than his *The Expression of the Emotions in
Man and Animals* (1872).

Is it corrosive to argue that the gestures, grimaces, modes of
expressing emotion are the same all over the world? For Darwin the connection of body and "spirit"—the complex of feeling and thought that constitutes us all—is a given. He cites
Claude Bernard on the way in which actions of the mind have
immediate effects on the heart and the actions of the heart reverberate back into the mind—the physical and the mental absolutely connected.

It is certainly this aspect of Darwinian thought, exploited and
developed most elaborately these days by evolutionary psychology, that seems most threatening to those who seek Dennett's despised "sky hooks," and that threatens most forcefully to disenchant the world. It is this aspect that manifests itself in Darwin's
argument in the *Descent* that our moral and aesthetic tendencies
derive ultimately from the work of natural selection, out of sexual drives and herd behavior. Darwin is, indeed, an exponent of
the view that there is a single human nature, an argument central
to much of the work of evolutionary psychology. But there is a
single living nature, as well, one in which the corporeal and the
"spiritual" are entirely integrated, the former the source of the latter. And, consistent with my argument about one of secularism's

most important qualities, the recognition of that integration and even of that priority is not at all demeaning, from Darwin's point of view. The trick is to allow oneself to value the corporeal, to recognize it not as a means but as an end, and to feel the wonder of its power to produce feeling and intelligence.

In *The Expression of the Emotions* Darwin moves easily from the way ducks behave to the way humans behave, allowing nothing in his prose to suggest that anything denigrating has been said. The continuity of animals with humans is part of the book's continuing argument, as is the continuity of humans with humans. Implicit in this leveling is the commitment Darwin makes elsewhere, most particularly, as Desmond and Moore claim, in *The Descent of Man* (of which *Expression* was originally intended to be a part) to the view that, as the Wedgwood cameo put it, "Am I not a man and a brother?" "The unity of the human races remained central to his science," claim Desmond and Moore (*Descent*, xviii). All races are bound together within the species human. Darwin's hostility to racism was persistent and strong, although in his own way he shared the basic Victorian sense that some races were lower than others, and, of course, that women were intellectually and physically inferior to men. *Expression* does not make a fuss about any of these ideas: its business is simply to trace universal traits by which all of the animal kingdom and all of humanity express emotions, but Darwin knew that this project would be "corrosive" to established Anglican authorities, though liberating to antislavery folk like him and his fellow Whigs.

Every sentence, then, however dispassionate and empirically convincing, is loaded with *affect*, but affect that it behooved Darwin to keep as quiet as possible. Yet as *Expression* is sustained primarily by anecdotal evidence, it carries with it an enormous weight of human interest that it is impossible to overlook, while it also demonstrates Darwin's familiarity with the arts. But perhaps most interesting of all for my purposes here is that it suggests the most nuanced attention to human and animal feelings, a very moving attentiveness in which science once again shows itself to be intricately enwound with the complications and emotions of everyday life.

Darwin, who shortly after the death of Annie returned doggedly to work on his cirripedes, wrote this in his clinical discussion of the manifestations of grief:

> When a mother suddenly loses her child, sometimes she is frantic with grief, and must be considered to be in an excited state; she walks wildly about, tears her hair or clothes, and wrings her hands. This latter action is perhaps due to the principle of antithesis, betraying an inward sense of helplessness and that nothing can be done. The other wild and violent movements may be in part explained by the relief experienced through muscular exertion, and in part by the undirected overflow of nerve-force from the excited sensorium. But under the sudden loss of a beloved person, one of the first and commonest thoughts which occurs, is that something more might have been done to save the lost one. . . .
>
> As soon as the sufferer is fully conscious that nothing can be done, despair or deep sorrow takes the place of frantic grief. The sufferer sits motionless, or gently rocks to and fro; the circulation becomes languid; respiration is almost forgotten, and deep sighs are drawn. All this reacts on the brain, and prostration soon follows with collapsed muscles and dulled eyes. An associated habit no longer prompts the sufferer to action, he is urged by his friends to voluntary exertion, and not to give way to silent, motionless, grief. Exertion stimulates the heart, and this reacts on the brain, and aids the mind to bear its heavy load. (*Expression*, 84–85)

Here again, Darwin makes his science out of imaginative projection of himself into his subject—in *Descent*, the Argus pheasant; here, a man who could stand in for Darwin in a piece of autobiography. Standing outside the condition he describes, however, he explains with remarkable attentiveness and precision, not only the sufferer's expressions, but the cause of them—which are not visible. Here again, then, he is both participant and dispassionate observer, and in drawing on his own experience and in connecting the condition of mind and spirit to the body, he does not diminish them. Darwin missed nothing, from the arching of dogs' backs approaching possible combat, to kittens kneading at their mother's mammary glands, to horses "nibbling" when stroked as they nibble to scratch themselves, to the raising of eyebrows in surprise, the retraction of lips in anger, and the protrusion of lips in sulkiness.

Darwin proceeds through the many various aspects of human feeling and the connection of their expression to bodily movements (often tied to survival strategies and therefore to natural selection), and all the while he maintains the detached observer's position as he is providing the evidence that a whole world of spirit, as it was traditionally understood, is grounded deeply in the corporeal. But there is nothing overtly iconoclastic in the language, unless it is Darwin's really daring combination of almost novelistic attention to states of feeling, to conditions of sadness, with scientific attention to physical conditions.

Here, for example, is an anecdote he uses to help explain certain widely recognized expressions of contempt:

> I never saw disgust more plainly expressed than in the face of one of my infants at the age of five months, when, for the first time, some cold water, and again a month afterwards, when a piece of ripe cherry was put into his mouth. This was shown by the lips and whole mouth assuming a shape which allowed the contents to run or fall quickly out; the tongue being likewise protruded. These movements were accompanied by a little shudder. It was all the more comical, as I doubt whether the child felt real disgust—the eyes and forehead expressing much surprise and consideration. The protrusion of the tongue in letting a nasty object fall out of the mouth, may explain how it is that lolling out the tongue universally serves as a sign of contempt and hatred. (159)

There is just a touch of evidence that Darwin felt for the baby in his apparent need to assure us that the child is not really disgusted (could this also be paternal self-excuse?). But of course, the attention itself implies feeling, the care with which Darwin registers the movement of the mouth and notices how the eyes and forehead express "surprise and consideration," and then speculates on whether the child "felt real disgust." The recognition that feeling is attached to bodily instinct in Darwin is virtually always inspiriting, surprising, almost fun. Value does not go out of the world as feelings are set inside the continuing struggles of natural selection. One of the most characteristic responses to

Darwin's registration of modes of feeling is the delighted surprise a reader is likely to feel that Darwin has got it right, that
that's the way it happens. (I chose this passage among hundreds
of others because it applies so immediately to a recent visit with
my granddaughter, at which I witnessed her first surprised encounter with feta cheese. There was Darwin's baby, and there
I'm sure was Darwin, as I found myself laughing sympathetically at her response, her face unmistakably expressing shock
and disgust as the "lolling tongue" tried to get that awful stuff
out of her mouth.) So the disenchanting rethinking of the expression of emotions provides moments of enchantment to anyone who has watched a baby or a cat or a dog or a horse or a duck
or primates in the zoo or actors at work.

And everywhere in Darwin there emerges a sense of wonder,
although it is not unself-conscious and not, certainly, his first objective. It would be simply disingenuous not to attend to the way
Darwin consciously attempts to secularize "wonder," most often
his own, though frequently generalized in a form that suggests
that everyone would be likely to feel it in relation to the phenomenon he is discussing. By that I mean, he wants to find a way to
demystify it after he registers it, precisely by subjecting it to careful consideration from his scientific point of view. In this respect,
he obviously follows the formula that Weber developed to explain disenchantment. The world is disenchanted precisely because of the way science, in principle, suggests the possibility of
explaining everything. Darwin wants to explain everything. But
clearly he feels no disenchantment, no diminishment in the thing
just *because* he can explain it. There is, as it were, an analogical or
metaphorical imagination at work that allows Darwin both to
particularize things brilliantly and to find their analogues: the
baby spitting out a cherry evokes for him all acts of contempt
that are expressed facially by "representing the rejection or exclusion of some real object which we dislike or abhor" (260).

His effort to explain leaves space for the wonder anyway, a
wonder no longer simply awed and dumb in the face of mystery,
but a wonder nevertheless. Moreover, most often in the course

of his "explanation," he produces the sort of argument that is it-self "wonderful," in that the mechanism that explains the mystery is itself virtually sublime in its working. Geological time, after all, though it allows us to credit Darwin's argument about the way species develop, is so awesomely beyond the imagination of human consciousness that it must even in its function as ex-planation induce something like the feeling of the sublime. As Beer has suggested, Darwin relies often on metaphors that some-times reach stunning complexity and richness. The metaphors in a way make the argument possible and are clearly an unskim-mable aspect of the thought that got Darwin where he wanted to be. Metaphor often allows Darwin to conceptualize processes that are indeed sublime.

This strange Darwinian secularization of wonder is registered, for example, in a long and metaphorical passage near the end of *Variation of Animals and Plants under Domestication* (1868), after a summary of his point that all races, genera, and species are "the descendants of one common progenitor," having developed "from simple variability."

> To consider the subject under this point of view is enough to strike one dumb with amazement. But our amazement ought to be lessened when we reflect that beings almost infinite in number, during an almost infinite lapse of time, have often had their whole organization rendered in some de-gree plastic, and that each slight modification of structure which was in any way beneficial under excessively complex conditions of life has been pre-served, whilst each which was in any way injurious has been rigorously de-stroyed. And the long-continued accumulation of beneficial variations will infallibly have led to structures as diversified, as beautifully adapted for various purposes and as excellently co-ordinated, as we see in the animals and plants around. Hence I have spoken of selection as the paramount power, whether applied by man to the formation of domestic breeds, or by nature to the production of species. I may recur to the metaphor given in a former chapter: if an architect were to rear a noble and commodious edifice, without the use of cut stone, by selecting from the fragments at the base of a precipice wedge-formed stones for his arches, elongated stones for his lin-tels, and flat stones for his roof, we should admire his skill and regard him as the paramount power. Now, the fragments of stone, though indispensable to the architect, bear to the edifice built by him the same relation which the

fluctuating variations of organic beings bear to the varied and admirable
structures ultimately acquired by their modified descendants. (*Variation*,
2:413–14)

The paragraph begins, then, with the most controversial, disturbing, and central point of all of his work after the *Origin*: that
all organic beings are "descendants of one common progenitor."
Characteristically, Darwin registers the shock that he knows his
readers will feel and that we can reasonably infer he feels
himself—it is "enough to strike one dumb with amazement."
The work of *Variation* is precisely to demonstrate, through study
of variations under domestication, that this is the case, in effect
to "domesticate" a culturally devastating conclusion. The affect
of the passage begins in noting how the fact might strike us
dumb, but, it would seem, Darwin wants to calm us down. The
big effect—and this is the central mode of his gradualist arguments from the beginning—is the product of simple events that
everyone might recognize, simply multiplied over vast periods
of time.

But the explanation is itself enough to strike us dumb, first,
because so important and massive an effect is a product of such
a "simple" cause, but more importantly because Darwin's explanation entails the invocation of a whole set of other staggering
facts, the most striking being geological time. Our amazement,
Darwin tells us, "ought to be lessened when we reflect" on this
process. But it should be lessened by a compilation of "infin-
ites," each one of which violates what had long been culturally
dominant assumptions about the nature of life, although Darwin was working within a tradition most prominently developed by Hutton and Lyell. To achieve the current conditions of
life, Darwin invokes "beings almost infinite in number," "an almost infinite lapse of time," and "excessively complex conditions." The vista that is to lessen our amazement is sublime in its
complexity and multiplicity and extension through time.

Darwin seems to understand that the breathtaking vision, com-
pressed now in a final paragraph, needs to be domesticated yet
further through a metaphor. Yet the work of metaphor, even if its

strategy is to familiarize, moves the subject invariably from the ostensibly dispassionate discourse of fact into a range of experience laden with value and feeling. Darwin's narrative, in which he is, in effect, building the metaphor of "selection" by using domestic selection as the primary evidence for natural selection, entails his need to demonstrate that in nature the work of "selection" can be done in a non-Paleyan way, not by intelligent design, or God, but by "long-continued accumulation." But it is exactly that mode of selection that Darwin recognizes is under potential attack. How can mere time and accumulation produce the sorts of effects that lead to "structures as diversified, as beautifully adapted for various purposes and as excellently coordinated, as we see in the animals and plants around us." Here, as in the *Origin*, the most important "animal," the one whose presence is really the reason Darwin requires such complicated rhetoric, is not specifically mentioned.

The "architect," instead of having stone cut to fit the needs of his plan, has to work with the stones available—as nature works with the variations given. In this respect, "design" is based on chance, on what's available among the stones that, through millennia, the weather, erosion, earthquakes, and other natural catastrophes have produced. In the following paragraph, Darwin deals with this problem through an extension of the metaphor, invoking "a savage utterly ignorant of the art of building," who will not, on the basis of explanations of *how* the stones were being used in the structure, understand what produced the stones in their peculiar shapes in the first place. So, the analogy then goes, are we ignorant in relation to the variations and differences in the development of organic beings. We can see *how* they work. We cannot know the "cause of each individual difference in the structure of each being" (414).

Darwin is, as ever, in the midst of a complicated argument, like the "one long argument" of the *Origin*. But it is an argument that can never shake the powerful feelings, starting with "dumb amazement," that anyone watching the phenomena must feel. Darwin's work is not to expel the feeling but to shift its ground. Paley asks for wonder at the astonishing working of God; Darwin, through his metaphors, seeks to shift the wonder to the astonishing work of nature. We can only "admire the skill" of the architect, who uses the chancy stones to produce a "noble and commodious edifice." What drives the metaphor is the sense Darwin tries to communicate that nature has the selective power that he dramatizes in the architect. Particularly in the paragraph that addresses the inadequacies of the "savage's" relation to the architect's work, we see that Darwin understands how the whole experience is laden with value and feeling.

The irony of this passage is that—as we have seen in Darwin's treatment of sexual selection—it echoes the metaphorical strategy of William Paley in *Natural Theology*, which is the argument that noble edifices and admirable structures can only be designed and that design requires a designer. Darwin's primary work is to eliminate the designer, but from the start the metaphorical presence of a designer in "natural selection" had been inescapable. Darwin wants to have it both ways: yes, the shape of stones is "accidental," but "this is not strictly correct" (414). Not correct, because the shape of each rock "depends on a long sequence of events, all obeying natural laws." *Variation* in fact ends by evading the large metaphysical questions Darwin's ambivalence produces, for he says, "here we are led to face a great difficulty in alluding to which I am aware that I am traveling beyond my proper province." As at the end of the *Origin*, Darwin here invokes an "omniscient Creator," but he asks, "can it be reasonably maintained that the Creator intentionally ordered . . . that certain fragments of rock should assume certain shapes so that the builder might erect his edifice?" And then he proceeds with something like passion to ask this creator further about how he could have foreordained the "many injurious deviations of structure as well as the redundant power of reproduction which inevitably leads to a struggle for existence, and as a consequence, to the natural selection or survival of the fittest" (415). The last line of the two long volumes that have meticulously traced the ways in which domestic variation works falls into a characteristic Victorian agnosticism: "Thus we are brought face to face with a difficulty as insoluble as is that of free will and predestination."

Throughout this whole long, difficult sequence, Darwin contends not only with the literal complications of his subject but with the religious and philosophical traditions to which, implicitly, his new scientific myth of origins is opposed. The stakes are high, and nowhere along the way does Darwin dispel the powerful affect that accompanied his attitudes toward nature right from the start, or that, as he understands, inhere in the questions he raises about the meaning, order, and origins of the world. Darwin's anger at the failures of the world that Paley saw as designed by a beneficent Creator is manifest throughout: would God, he said, "ordain that the crop and tail-feathers of the pigeon should vary in order that the fancier might make his grotesque pouter and fantail breeds? Did he cause the frame and mental qualities of the dog to vary in order that a breed might be formed of indomitable ferocity with jaws fitted to pin down the bull for man's brutal sport?"

The passion in Darwin's relation to nature, moving from dumb amazement at the derivation of all organic beings from a single one, to confrontation with the sublime prospects of geological time, to awe at the extraordinary multiplicity and diversity of organic life, includes horror at so much that goes awry, so much that is distorted, cruel, violent. His resentment against the beneficent, omniscient Creator who might be thought to have produced such horrors is deep.

But while Darwin's vision includes a recognition of mindless horrors (better, he thought, than mindful ones), and while its very strategies of reexplanation, entailing as they do a move to see magnificent things as the product of small incremental changes, might be taken as "corrosive," in the long passage we have been examining, Darwin describes the natural world as an object of awe, wonder, and terror. Moreover, while theoretically one might explain the causes that produced those rocks the architect needs to use, they remain in fact, and will always remain, unexplained. Darwin looks at this world not as something dead and remote from human values but as fraught with those values. His growing hostility to the Creator, or to those who insist on the Creator's work in this strange, awesome, dangerous world, has to do with his deep sense that as we approach nature, we have every right to regard it with the moral vision of his time. He is appalled by man's "brutal sport," but he is alert to the way in which so many variations within nature are "of no service to man," and are often "injurious to the creatures themselves."

The sort of enchantment, then, that one might feel in Darwin's work is not a mere cooing at how pretty the world is, or "dumb amazement" at its sublime extent and variety; it has some of the chilling effect of the sublime itself. It is merely mawkish and sentimental not to notice that while we are enchanted by this world, "struck and shaken by the extraordinary that lives amid the familiar and the everyday" (Bennett, 4), awful things are happening, and often to people who have no space in which to feel the enchantment. There is a famous passage in the *Origin* that can capture something of the doubleness of this enchantment narrative that I am asking that we derive from Darwin:

> We behold the face of nature bright with gladness, we often see superabundance of food; we do not see, or we forget, that the birds which are idly singing round us mostly live on insects or seeds, and are thus constantly destroying life; or we forget how largely these songsters or their eggs, or their nestlings, are destroyed by birds and beasts of prey; we do not always bear in mind that though food may be now superabundant, it is not so at all seasons of each recurring year. (62)

The Darwinian vision takes us beyond what we can literally see, although nobody was more determined to be empirical in his arguments. His theory is both counterintuitive and dependent on a deep imagination of movement and changes in time, of ambient worlds that, while not readily seen at any given moment, account importantly for what we can see, and of others and otherness. It is a strenuous, difficult, sublime world; it is a world full of terrible things, like the death of Darwin's daughter Annie, or the deaths of all those creatures less fit than others for reproduction and survival. It is the world we live in, and it allows for joyful and painful attachment.

But let me here turn back to an aspect of Darwin we have already glimpsed in the passage from *Expression*, in which he recounts his child's rejection of cold water and cherries. His empiricist determination to remove himself from his work is suggested also by his "Observations on Children," a journal that recorded in detail his children's behavior and speech and later provided materials for such works as *The Expression of the Emotions in Man and Animals*. As the editors of the ongoing edition of Darwin's correspondence put it, Darwin possessed the ability to dissociate himself sufficiently from his own emotions to "make use of the most readily available resources—in this case his own children—for information on emotional expression, just as he had earlier analyzed his own childhood memories" (in the "autobiographical fragments" of 1838) (*Correspondence*, 4:410). I will later on touch briefly on his journal about his first child, William, which may usefully suggest how this apparent detachment of Darwin's is compatible with the deepest of family feelings, and by projection, how Darwin's intense observations of the natural world reveal "a feeling for the organism." But at first glance, it would seem an almost heartless business to turn one's own children into scientific subjects. This is an aspect of Darwin that has to be confronted: is his eagerness to make all aspects of his life, personal and other, grist for his scientific mill evidence of heartlessness or of a sense of a natural world thick with value and feeling?

Darwin, as we know, kept a journal recording the behavior of his first son, William.[15] He began the journal, as John Bowlby points out, because he thought that if natural selection really were in operation, one might detect in the unreflective behavior of infants the inheritances and developments that would carry survival value (243–44). So his journal reveals more than a touch of impersonality in its attitude toward this first child of his.

Published in *Mind* in 1877 (forty years after it was written) under the title "A Biographical Sketch of an Infant,"[16] Darwin's essay seems merely descriptive. It registers no overt affect on the observer's part, though it fascinatingly suggests a relationship not apparently relevant to the materials under discussion. For example, once, when William was sixty-six days old, Darwin happened to sneeze, and William "started violently, frowned, looked frightened, and cried rather badly: for an hour afterwards he was in a state which would be called nervous in an older person, for every slight noise made him start" (465). Darwin could pay such attention to barnacles, worms, and ants as well, but William, of course, was not just another living thing. Darwin's attentiveness is both paternal and scientific, the paternal transformed, for the purposes of the journal, into the scientific but bearing with it that observant lovingness that is implicit in Darwin's work throughout, and that perhaps most painfully one finds in his letters to Emma during Annie's last days. Something of Darwin's feeling for the baby whose progress he so scientifically recorded can be inferred from a letter he wrote to his good friend W. D. Fox: "I defy anybody to flatter us on our baby,—for I defy anyone to say anything, in its praise, of which we are not fully conscious. . . . *I had not the smallest conception there was so much in a five month baby.*—You will perceive by this that I have a fine degree of paternal fervour" (*Correspondence*, 2:269). But Darwin's biographical sketch of an infant concludes undramatically, if tellingly, in a discussion of infant communication. "Before he was a year old, he understood intonations and gestures, as well as several words and short sentences. He understood one word, namely, his nurse's name, exactly five months before he invented his first word mum; and this is what might have been expected, as we know that the lower animals easily learn to understand spoken words."

One might even want to say—I do, in fact, want to say it—that this kind of almost detached attention can imply lovingness as subtly as scientifically descriptive language has been taken to imply cold detachment. The famous picture of Darwin, a gentle smile on his face, with William on his lap, shows us nothing of the detached, disinterested investigator using his son as a piece of evidence in the construction of his theory. Yet the journal is scrupulous about the details even as Darwin remarks on how William, "after grasping my finger and drawing it to his mouth," finally learned how, on his 114th day, "to get the end of my finger *into* his mouth" (Gruber, 467). He claims scientifically that William first manifested shyness when he was nearly two years and three months old. This, Darwin notes, was "shown toward myself, after an absence of ten days from home, chiefly by his eyes being kept slightly averted from mine; but he soon came and sat on my knee and kissed me, and all trace of shyness disappeared" (472). Science does not disguise the relationship. In his correspondence, he asks Emma to give "Mr Hoddy Doddy a kiss for me" (2:263). It is pretty clear that William, his first infant subject, is an object of love.

The loving and scientific act is particularly difficult because it threatens to destabilize the conditions it has been seeking to understand and sustain. I hear Darwin described, even as I resist the description, in Pratt's painful and accusatory description of the eighteenth-century natural historian: "a European bourgeois subject simultaneously innocent and imperial, asserting a harmless hegemonic vision that installs no apparatus of domination" (33–34). Certainly, Darwin's interest in tracing the development of his child marks an important moment in the positivist pursuit of a human science, forwarding the ambition that James Mill had announced to make the workings of the mind as clear as the road to Charing Cross. And such positivist efforts have emerged with a distinct biological and Darwinian twist now in evolutionary psychology. Darwin is inside that track. But he is also outside of it.

There is no escaping one's own presence—that's one of the lessons of most recent epistemology; but, ironically, it is also one of the implications of political critique of presence. On the one hand, that is, whatever disinterested gestures the scientist and naturalist might make, those gestures always disguise a potentially disturbing ideological intrusion. On the other, however, though it is not usually overtly asserted, those who criticize the aspiration to unmarked interest or the claim to have achieved it tend to imply through the critique that the best thing for the truth would be to eliminate human presence and human activity completely. To criticize objectivity because it does not attain absolute disinterest is at least in part to imply that absolute disinterest would be a good thing—if we could get it. Radical distrust of the work of acquiring knowledge subjects us often to a kind of epistemological hypochondria. Let's agree that there is nothing for it; every human activity has implications it doesn't consider, and certainly Darwin's enterprise, so brilliantly and exhaustively historicized as it has been, needs to be read inside a cultural and ideological history about which Darwin was not particularly reflective. But within the problematic created by this historicizing, the alternatives cannot be the false ideal of absolute rationalist disinterest (a strategy that a quarter-century of cultural theory and philosophy has demonstrated to be unattainable) or the false ideal of full ideology critique. It is possible, and I have been arguing, necessary, that the detached scientist have a passionate engagement with his or her materials. It might be thought almost a condition for good science that the scientist have a deep sense of the value of the things being studied and of the study itself. A feeling for the organism is a condition for understanding how that organism works. Once past the dichotomy between complete detachment and ideological complicity, it should be possible to recognize that a world explained by science can be enchanted after all, invested with a powerful imaginative sympathy, potentially explicable, yes, but awesome and astonishing in its operations.

The false vanishing acts of absolute disinterest are not adequate alternatives. The world can be both rationally described and ideologically attractive; seekers of knowledge can attempt both to efface their interests and feelings in the pursuit and to care deeply about the world they are discovering. Darwin's language, working toward one of the great imaginative and intellectual achievements of modern times, implies another possibility, the very human and *necessary* presence of the Weberian explainer in the workings of nature. Darwin's study of his children can be taken as a kind of metaphor for a science that leaves the world thick with value, meaning, and affection. It implies a relation to nature not dualistic but utterly engaged, entwined, loving: the layer that cannot be skimmed off in the quest for the pure algorithm.

To the modern lay reader, perhaps the most striking element in Darwin's meticulous observations and descriptions of natural phenomena and his strenuously constructed mind experiments and argumentation is its ostensibly unscientific exuberance. Although the structure of his argument entails a determined refusal of any kind of nonnatural intervention, Darwin is stunned by the extraordinary variety and beauty of what he sees. It is not beside the point that the concordance to the first edition of the *Origin* lists twenty-nine entries for variations on the word "beauty," forty-two for "wonderful," and fifteen for "marvelous." Perhaps equally important for the overall effect of the prose, there are fifty-seven "unknowns." That so much is unknown and yet to be discovered only increases the sense of marvel and wonder.

Darwin's effectiveness depends partly on his extreme sensitivity to what would be perceived as strange and marvelous and incredible. For him, the wondrous and the marvelous needed explanation within the terms of secondary causes, but such explanation did nothing to diminish the wonder. If they could not be explained they would have been, Darwin said at least six different times, "fatal" to his theory and his views. Thus, the extensive sections in the *Origin* called "Difficulties on Theory" were possible in part because Darwin recognized, not only scientifically but viscerally, what needed to be explained away. The disenchanting rationalist move depends, at least in part, on the romantic intensity of his responses to organic phenomena, his power to imagine himself into others, and into the sensibility of his readers.

There is a famous and curious reversal in Darwin's discussion of the eye, a phenomenon whose ostensible perfection seemed to be a particular obstacle to the theory of natural selection, not least because Paley himself had focused on the eye as one of the most striking evidences of manifest divine intention. Reason, in Darwin's argument, requires the fullest sort of imagination. In a passage quoted in a previous chapter, it will be remembered, he effectively turns reason and imagination on their heads, evoking reason as a power more risk-taking than imagination to overcome the difficulties on the way to recognizing that "a structure even as perfect as the eye of an eagle" might be the product of natural selection.

The distinctly Darwinian note here, aside from the marvelous irony of the inversion, is Darwin's recognition of how "startling" a naturalistic description of the extraordinary phenomenon of the eagle's eye really is. Darwin's passion for reason is deeply imaginative, for the reason invoked entails something like a leap of faith, and the imagination is constrained by the merely credible. What Darwin's experience of nature produces is a sense of the incredibility of what he is trying to argue—the theory *is* largely counterintuitive. Paley is "imaginative," and Darwin is "rational."

Each of Darwin's defenses of his theory bears the weight of two conventionally opposed elements. On the one hand, the point of each argument, as with that of the eagle, is to demonstrate that explanation through "secondary causes" is possible. It is the corrosive algorithmic description that Dennett so admires and exploits. On the other hand, Darwin is deeply sensitive to the fact that any normal perception of the phenomenon will turn first to a nonnaturalistic explanation, be it the eye of the eagle, the tail of the peacock, or in the end the mind of man. "Instinct," he recognized, was a particular obstacle. It is one thing to manipulate the length and shape of a pigeon's tail; it is quite another to account for "so wonderful an instinct as that of the hive-bee making its cells" (*Origin*, 207), or the migration and nesting habits of cuckoos, or the slave-making behavior of certain ants.

Darwin begins his stunning, indeed virtuoso, chapter on instinct with an immediate recognition of how "it will have
occurred to many readers" that instinct is a subject of "difficulty
sufficient to overthrow my whole theory" (207). The Darwinian
strategy here, entirely characteristic, is to imagine the position of
the reader, and he can do so by recognizing, and indeed feeling,
the wonderful nature of his subject. Throughout the chapter, he
firmly insists that natural selection can produce wonders but
makes the case by pursuing details strenuously (apologizing on
the way for going on so long about the details of hive making, for
example, and at the same time for not taking as much time as
necessary). The enthusiasm and admiration that Paley would
have turned to God, Darwin turns to natural selection and the
development "from a very few simple instincts" (224). The end-
ing of the chapter on instincts has a kind of bold and indeed un-
sentimental toughness that moves toward the corrosiveness Dennett affirms. It is in fact an unusual sentence in its boldness. But
interestingly, Darwin slips back into the language of imagination rather than the reason he claims must replace imagination:

> Finally, it may not be a logical deduction, but to my imagination it is far
> more satisfactory to look at such instincts as the young cuckoo ejecting its
> foster-brother,—ants making slaves,—the larvae of ichneumonidae feeding
> within the bodies of caterpillars,—not as specially endowed or created in-
> stincts, but as small consequences of one general law, leading to the ad-
> vancement of all organic beings, namely, multiply, vary, let the strongest
> live, and the weakest die. (244)

Tough as this is, it is also—especially following the extraordinary arguments and details of the chapter—rich with a sense of
the marvelous and at last gives credit where it is due, to the
work of imagination.

Even with this relatively unusual macho rhetoric, natural
selection—the book's central metaphor—is unskimmably anthropomorphized throughout the first edition. It may simply disguise an algorithm, but Darwin's rhetoric concerning it circles
around the word "good." It works for the "good" of the creature
it tends, never, however, "for the good of another species" (87). It

is always "working for the good of each being" (194). It will "never produce in a being anything injurious to itself, for natural selection acts solely by and for the good of each" (201). It is clearly maternal (and implicitly loving of each creature in its care), and it replaces Paley's God, whose literal intention determines the shape of things, with a wonderfully attentive, if tough, love that cares for every hair of the head. Natural selection, in the book that employs it to banish intention from the workings of nature, is the extraordinarily complex protagonist of a narrative that evokes wonder in every detail, for—in the most extended and most famous passage about it—natural selection

> is daily and hourly scrutinising, throughout the world, every variation, even the slightest; rejecting that which is bad, preserving and adding up all that is good; silently and insensibly working, whenever and wherever opportunity offers, at the improvement of each organic being in relation to its organic and inorganic conditions of life. (84)

Despite the strong tradition naming itself Darwinian that emphasizes and even enjoys the heartlessness and mindlessness of natural selection, Darwin's language consistently imagines it differently. Skimming away the personification (as Darwin himself often felt obliged to do) might justify the painful interpretation, but Darwin, the messenger of the bad news, remains awed by the workings of natural selection, and deeply admiring. In the third edition, Darwin notoriously defends his metaphor, natural selection, by claiming that

> Every one knows what is meant and is implied by such metaphorical expressions; and they are almost necessary for brevity. So again it is difficult to avoid personifying the word Nature; but I mean by Nature, only the aggregate action and product of many natural laws, and by laws the sequence of events as ascertained by us. With a little familiarity such superficial objections will be forgotten.[17]

Here, it seems, Darwin is doing his own skimming, defensively suggesting that the metaphor is only a convenience, a dressing— necessary for "brevity." But the move to rational disenchantment of what seems like—on Darwin's own account—"an active power of Deity" is at best disingenuous. Not everyone to this day "knows what is meant." The metaphor was clearly critical to Darwin, as he developed his theory on the basis of and in contrast to artificial selection. "Natural" selection both contrasts with and parallels artificial selection; in Darwin's world, nature does, in some quite literal way, select, although without consciousness.

Moreover, as we read through Darwin's work, we find that natural selection and then sexual selection, with obviously more real "intention," has peculiarly human responses to things. It is reasonable to ask whether the "selection" in sexual selection is the deliberate activity of the organisms involved, or a large impersonal and mindless process operating through them. The organisms that do the selecting (at least the birds and the mammals) are susceptible to the beautiful. Darwin talks, for example, of male birds "displaying their gorgeous plumage, and performing the strangest antics before an assembled body of females." "We cannot doubt," he claims, "that though led by instinct, they know what they are about, and consciously exert their mental and bodily powers" (1:258). He describes, as well," the gorgeous butterflies that inhabit the tropics," the "beautiful and curious mustache monkey" (2:291), "the strange habits in certain species" (*Origin*, 212), the "beauty and infinite complexity of the coadaptations between all organic beings" (*Origin*, 109).

Despite that inadequate passage in the chapter "Struggle for Existence"—it isn't even true that "the war of nature is not incessant," that "no fear is felt" by the victim, and that "death is generally prompt" (79)—we have seen that the idea had no consolatory power for him after his daughter's death either. It is true, however, that the chapter does more than enumerate struggles; it includes in the idea of "struggle" what looks a lot like love, a point that Kropotkin emphasized. But more important, there is no weakness in the work's exuberance. It makes manifest Darwin's own deep concern, paralleling that of his protagonist, natural selection, for every minutest detail of organic life and his profoundly attentive and loving notation of organic life (he claimed, remember, that one of his few special virtues as a scientist was his power of observation). That his theory tells us some very bad news about the world's indifference to us and the material connections between us and all other organisms is only half of the story. His world is nonetheless full of "grandeur." Yes, like most other naturalists he characterized nature as a she, and, yes, he could make his discoveries, his arguments, and his metaphors because he could live off things like railroad shares and pay to join an expedition aimed at mapping the world's coastline for the British admiralty. It would be merely disingenuous to suggest that he was not part of that imperial tradition Pratt describes, and yet Darwin's Romantic materialism insists on organic rather than mechanical metaphors and does not intrinsically imply the work of domination.

Dennett's algorithm strips the organic metaphors away—a logically legitimate but imaginatively diminishing strategy—while Darwin's natural selection remains persistently maternal, even after his own disclaimers. The Enlightenment heritage in Darwin moves to no necessary emptying of the world, no inevitable drive toward domination. And as to the question of enchantment, Darwin's argument does not explain the world away; it makes the world thrilling in its intricacy and richness and beauty, and it makes the experience of the world, now explained (or potentially explainable), more fascinating, more interesting, more beautiful. For me at least, it makes the world both more intelligible and more wonderful—there is, after all, grandeur in this view of things.

I have been trying to extract from the complex of Darwin's contingent mid-Victorian being, from his way of thinking and writing, a vision that is both scientifically valuable and explanatory and that allows an enchanted view of the world—or at least allows those moments of enchantment that Bennett describes. I have been trying to project a view that allows for an entirely secular, a naturalistically explicable sort of place, one that is worth caring for on its own terms, that is, in fact, too gloriously rich and alive *not* to care for. And I have wanted through Darwin to reexamine the taken-for-granted notion that the scientist is something of a science-fiction monster, so committed to knowledge that he (even she, though less often) will risk loosing a herd of mad, man-eating dinosaurs on the world. The division between reason and imagination needs to be worried through yet again. Darwin self-consciously confused them; his work demonstrates how much his good ideas depended on his being the person who he was, with the prejudices he bore.

This enchantment narrative that I am trying to write through Darwin draws on the philosophical tradition that William James has exemplified, that, as David Wilson has recently argued in a very Darwinian context, strict rational discourse does not inevitably trump feeling in every aspect of human relations (40–43, 228–29, passim). It is, says William Connolly, enchantment, the work of affect and the capacity for awe and reverence, that has been missing from Western secularism. And talking about religion as an adaptive mechanism, Wilson shows how nonrational mind work can be as humanly valuable, certainly in creating group allegiances, as rational argument. The point of a rationalist argument like Dennett's is that the most adaptive behavior is one that confronts the "truth," as good science will reveal it, no matter how painful and difficult.

One can rationally analyze the value of irrationality; one can recognize how central to Darwin's achievement were a set of Romantic passions and even bourgeois domestic values. Despite their failure—like the positivists'—to persuade the culture as a whole, many of the leading Victorian intellectuals were right to attempt to invest rational understanding of the world with affect, no matter how difficult the project. As I see it, it remains a critically important one. Watching Darwin maneuver through the development of his theory and his attempts to use that theory to explain more and more of the world, including, of course, human behavior, we can recognize that even while he was at work disenchanting, his language implied the enchantment of his subject. There can be no Darwinian religion, but there can be a spreading recognition that the most serious secular and rational engagement with the conditions of the world is not incompatible with the religious qualities that Wilson and Durkheim identify as essential to human survival.

I have taken Darwin as a model for the way science and the secular can inhabit an enchanted world. And Darwin's singular condition is not so singular. It may be easier for a rich man to be satisfied with a world entirely bereft of the possibilities of the transcendent, but my argument is that the attitude, the feeling, is available to virtually everyone. Even Dennett, I should concede, calls this secular world "sacred." Darwinian enchantment entails an attitude of awe and love toward the multiple forms of life, the recognition that on the diversity and the extraordinary variations that nature has developed through the expanses of geological time depends our own quite justified sense of sharp difference, but also our family resemblances, from fish to bat to human. Darwin's shock and horror at the first Fuegians he encountered on his *Beagle* voyage never disappeared, and yet, as Desmond and Moore show, his deep sense of racial connection pushed him to the arguments of the *Descent*, and his implicit insistence there that we all are indeed, in a very real, biological sense, brothers. In his pursuit not of divine explanation or moral justification but of the truth, he demonstrated the kinship with other human beings that at times appalled him (although he was entirely content to be related to the apes). Darwin could never come to love the Fuegians or to believe that they were evolutionarily the equal of his own race and class, but his "science" opened up that possibility, as it had opened up the possibility of female choice. It opened up as well the possibility of familial love as it recognized not spiritual but literal kinship. Darwin fully acknowledged and movingly acquiesced in the ways that nature might produce a diversity that disguised what many anthropocentrics didn't want to know. Darwin was indeed anthropomorphic in his thinking, but he was not anthropocentric: that is, while he projected with creative imagination his own assumptions and his culture's into the organisms he studied with such attention and love, he never allowed himself to assume that what they did, how they chose, how they had come to be, were somehow designed to satisfy human needs and desires. We are all related, yes, but the work of nature was not designed to satisfy us. The kinder, gentler Darwin I have been trying to describe achieves that kindness and that gentleness by refusing to impose his values and desires on nature; he uses those values and those desires, rather, to find out as much as he can about how we are all connected.

So it is a small first step Darwin offers us toward an enchantment that depends not on anthropocentric imaginations of the world but on anthropomorphic, imaginatively metaphorical impulses to understand it and love it, precisely in its refusal to be like us, whoever "we" may be. My son's bumper sticker is not an exaggeration or a joke: "Darwin loves you" after all.

EPILOGUE

What Does It Mean?

Cold comfort, perhaps, this unhallowed sacredness. That Darwin loved the world despite his illness, his losses, and all he knew may seem not to have much to do with our own particular conditions, most of us not scientists steeped in the particularities of nature. Scientists often take joy in a world that, as they describe it, may frighten and appall the rest of us. As the distinguished physicists George Charpak (a Nobelist) and Roland Omnès have written, "The universe and its laws evidently arouse strong feelings in researchers committed to their work. They derive great pleasure from it."[1] Darwin clearly did. This pleasure would seem to be singular and specialized, and whatever Darwin suffered when one thinks of the overall conditions of his life, it is difficult to imagine them as the norm for most of us. Certainly, Darwin had relatively little to complain about: a prosperous life that fairly rapidly brought him international fame, a lovely house, a totally dedicated and loving wife, many successful children. Most of us confront the strains of ordinary living and the spiritual emptiness that Weber described and lamented under much less supportive circumstances.

What if we have no income to speak of, no lovely house or dedicated spouse? The enchantment I have been invoking through the example of Darwin does nothing to justify the staggering inequities and disasters that characterize human society and the natural world, nor did I mean to suggest that it would. It would be merely absurd and entirely inhumane to insist that in the midst of poverty, brutality, suffering, or catastrophic natural disasters, like the recent tsunami or the hurricane that wiped out New Orleans, people wake up to the astonishing beauty and diversity of life. The world is a hard place, but it need not be disenchanted; wonder at the natural world does not disappear with poverty, nor with a belief that the world behaves according to "natural law," whatever that might mean, and that it might be possible for humans to investigate any given phenomenon with some hope of coming up with a satisfactory naturalistic explanation. The scientist's tendency to love the world he or she is committed to describe as impartially, with as little affect as possible, is not an aberration at all. It is what happens to any of us when, for one reason or other, we really learn to see something with great clarity, to understand its workings. It simply won't do to complain, as Jane Bennett has formulated the negative argument, "that only effete intellectuals have the luxury of feeling enchanted, whereas real people must cope with the real world" (10). Everyone has to cope with the "real world," and everyone can respond to its astonishing powers, its enormous complexity, its overwhelming diversity, and its remarkable vital energy. In the bare, tedious flatness of a shopping center parking lot, grass and weeds force themselves into the cracks. In the cavernous warehouse spaces of new superstores, house sparrows chirp and make their nests.

The Darwinian enchantment this book has been proposing and, perhaps, preaching has its problems, not least the very subjectivity and privacy of the experience as I have been describing it. The modern enchantment that Connolly and Bennett espouse, although it is conceived as a way into otherness and into a public life of caring and generosity, is likewise too private to satisfy some of the needs that organized religion satisfies. But religion survived the onslaught of scientific naturalism in the nineteenth century in part by going private and personal. There was, as Callum Brown points out, a vast resurgence of popular religion just when Darwin and then the scientific naturalists were spreading the new gospel of agnosticism.[2] That resurgence—child of Romanticism—was focused on feelings and the conduct that issues from them and was resistant to theology and institutionally organized religion. The loss here is of just those qualities of the religious life that depend on public rituals and what Charles Taylor calls "collective practice." Taylor points out that "a striking feature of the Western march toward secularity is that it has been interwoven from the start with this drive toward personal religion" (*Varieties*, 13). And I recognize in the secularism of nontheistic enchantment just those qualities of resurgent popular religion that mark it as personal, mark it as questing for authentic experience and resisting external forms of belief and ritual. The danger of nontheistic enchantment is similar to the dangers of the religion of personal experience and the claims to authenticity that go with it.

The critical issue is how to move from the "authentic" personal experience of nontheistic enchantment, from the intensity of personal engagement, wonder, awe, and love, to a less solipsistic, more public, more ethically engaged secularism that will bear within it that reverence for the world that is essential to an "ethically sufficient" life. The intellect, as Bennett argues (alluding here to Schiller) is insufficient for ethical action (140–43). The work of enchantment is partly to energize the ethical. I am aware that the Darwinian model shares the limits of an entirely personal religion, and as Bennett fears, might lapse into a mere aestheticism. It makes only a small first step toward a more humane and a more joyful, a more open, a more knowledgeable and attentive relation not only to the natural world but to the communities we inhabit.

I am aware too that the very argument I am making in this book, like the argument that Darwin made in his world-changing work, echoes with the rhythms, the strategies, the feeling of natural religion. It was the very feeling of wonder and awe that I claim is a condition of Darwinian enchantment that the natural theologians took as evidence of a supernatural creation. I was struck recently by an article in the *New York Times* about the reception of that remarkable and moving film documentary *March of the Penguins*. When I had seen the film, some weeks before, I was moved by its meticulous, brave, and understated depiction of how animals adapted for swimming and feeding in the water, but living in the cruelest climate in the world, managed to breed. Looking at the film through Darwinian eyes, I marveled at the tenacity and delicacy of the awkward and (anthropomorphically) charming penguins, at their strategies for survival, at the tenuousness of their existence, at the similarity in overwhelming difference that made their lives so fascinating and touching. I remember thinking when I came out how perilously the penguins' ability to survive was suspended on very particular conditions. And I immediately started on some little hypothetical mathematics of survival: how many eggs must each pair of penguins successfully hatch in order to sustain a population so vulnerable to the weather and predators and a way of life so austere and demanding. That these extraordinary birds had managed to survive in such brutal circumstances suggested to me, not the benign generosity of the natural world, but the delicacy of balance sustained by natural selection. It was clear that the minutest change in circumstances—perhaps a slight warming that would push breeding grounds further from the feeding grounds, even by only a mile or so—might mean the extinction of the species. The whole narrative seemed to me so powerfully and self-evidently a dramatization of Darwin's theory that it never occurred to me that the admiration and indeed awe that one comes to feel for those extraordinary creatures might have been taken as evidence that the evolutionary explanation won't do.

But then I found the *Times* reporting (to my initial shock and disbelief) that "At a conference for young Republicans, the editor of *National Review* urged participants to see the movie because it promoted monogamy. (That penguins change mates every year is a fact that the young Republican apparently missed.[3]) A widely circulated Christian magazine said it made "a strong case for intelligent design." It is true that, as the *Times* suggests, the movie tended to "soft-pedal" topics like evolution and global warming. But attention to anything but the cute waddling of baby penguins and the awkward tuxedo-like dress of the adult birds might have suggested something of what was really going on. "It's obvious that global warming has an impact on the reproduction of the penguins," Luc Jacquet, the director, told *National Geographic Online.* "But much of public opinion appears insensitive to the dangers of global warming. We have to find other ways to communicate to people about it, not just lecture them." There is, however, a key opening line, though it is only one, in which the narrator, Morgan Freeman, says that "For millions of years they have made their home on the darkest, driest, windiest, and coldest continent on earth. And they've done so pretty much alone."

But yes, even that seems translatable into the new natural theology, the theory of "intelligent design." The *Times* reported that "to Andrew Coffin, writing in the widely circulated Christian publication World Magazine," the difficulty [of the penguins' strategies of survival] makes "a winning argument for the theory that life is too complex to have arisen through random selection." It was a shock to realize that Coffin was responding to exactly the film I saw, the film that so strongly confirmed Darwin's theory and, at the same time, so strongly confirmed the power of the Darwinian vision to "enchant." Looking with non-Darwinian ideas beyond the extraordinary narrative of life here and now, Coffin seeks to make it a story of otherworldly presence and thus laments that "acknowledgment of a creator is absent in the examination of such strange and wonderful animals. But it's also a gap easily filled by family discussion after the film."

If this is "intelligent design," it is hard to imagine what unintelligent design might be. What designer with any competence and with any compassion at all would construct a mode of living and survival that entails so much pain, so much awkwardness, such clumsy reuse of organs and limbs apparently adapted for other purposes? Why force aquatic birds (with wings that don't work as means to flight but are already readapted for swimming) to "march" for seventy miles from their source of food to their breeding grounds, or to walk on their heels for months in order to protect the egg from touching the ice and immediately freezing? Was it an intelligent designer, or the penguins, who figured out that this was a manageable way to do things, and then did it? The genius of all this, the design, if that word must be used, seems entirely attributable to the birds themselves who have had to make do with rather bad design (given their circumstances). Clearly, their bodies were "designed" for other purposes than the ones they are fulfilling now. DNA shows that what the untutored eye would find it difficult at first to credit, that the penguins' closest living relatives are albatrosses, which not only have enormous wingspread, as much as eleven or twelve feet, but travel thousands of miles to feed and breed. The albatrosses' feats are almost as amazing as the penguins'. Both are remarkable, enchanting creatures. Both deserve to be admired, attended to, respected, seen for what they are.

The "intelligent design" advocates' short-sighted interpretation was, however, surpassed by the last evidence the *Times* supplied for the ways in which conservative groups have adopted the film and resisted the reality of the penguins themselves. To my combined horror and amusement, the story reported how a minister in Ohio read the film: "Some of the circumstances they experienced seemed to parallel those of Christians," he said of the penguins. "The penguin falling behind is like some Christians falling behind. The path changes every year, yet that they find their way, is like the Holy Spirit."[4]

It would be pointless to argue these issues here, but it is only reasonable to recognize the desperate need for "meaning," beyond what the natural world can provide, that drives these kinds of commentary. Since in this book, I have insisted on the possibility of nontheistic enchantment and on its importance, it is certainly necessary to recognize that my argument too is driven by feeling, and aims at engendering feeling in others. And so it is only reasonable to remind ourselves that both natural theology and nontheistic enchantment build on the William Jamesian insight that while the work of rationality and intelligence is indispensable, the intellect is not enough and will not by itself generate action and lived ethical commitment. Obviously, there is no way to argue that my feeling is somehow better than your feeling. But it is essential, first, to recognize that the commitment to a Darwinian and secular interpretation of such phenomena as the

march of the penguins is entirely compatible with those feelings
of wonder and awe that have traditionally been expressed by religion, and that commitment entails the fullest possible respect
and sympathy for the natural phenomena and the living beings
around us. Second, while resisting triumphalist rhetoric, we need
not surrender the commitment to turn mysteries into problems
for which we seek solutions, to reject those mystifications and
sky hooks that attempt to restore wonder by denying that mysteries have been turned into problems, after all. And third, it is
crucial that we follow the Darwinian model of precise and sympathetic observation, anthropomorphic, perhaps, but not anthropocentric (or theocentric for that matter). We owe it to the nature
we study to respect its difference, as Darwin most certainly did.
We might well try to think and imagine our way into the condition of those penguins without imposing on them the schema
and desires of a humanity grasping for consolation.

When, behind the Darwinian explanation, there has accumulated such an overwhelming mass of evidence that helps explain
life as we know it today and can, as it were, predict the past (or,
more precisely, what we can learn about the past) and even manipulate developments in the future, the feelings associated with
nontheistic enchantment seem validated. It is necessary to take
the risk, then, of using argumentative strategies parallel to those
that I have been resisting, just as Darwin used the strategies of
natural theology to deny it. Those strategies, which imply a
strong belief in the value and reality of enchantment, remain important today.

Bennett is right to insist that the myth of disenchantment,
like virtually all stories, is not only a story but is itself an actor
in the world. The assumptions that story embodies have their
consequences, and no assumption is more dangerous, more
devastating to the condition of living, than the view that nothing can matter unless it is somehow sanctioned by what is not
of this world—some telos to which it points, some God who
created it for some divine reason. This is an invitation to de-
spair and self-deception, and to the devaluing of every living

thing, of everything. Recharging the batteries of wonder by finding everything ultimately inexplicable, always a mystery, can make for a kind of dogmatic obscurantism that may seem to be valuing what is for what it is, but in effect values it for not being what it is, a natural entity of one sort or another. Post-Darwinian enchantment entails, on the contrary, an overwhelming sense of the deep value of everything, of the miraculousness of the natural order in every one of its manifestations—and obviously there are many such manifestations most of us would be happy to do without. Part of the mystery of nontheistic enchantment is precisely the world's diverse and radical difference from us, and its kinship with us. Darwin's major metaphor, "natural selection," embodies with great richness the paradox of a world entirely indifferent to each of us, and yet so structured that everything living in it—each of us included—is most scrupulously tended to by those indifferent natural processes: "Natural selection will never produce in a being anything injurious to itself, for natural selection acts solely by and for the good of each," he writes (201). But its "production" is full of astonishing "contrivances," and things that include what must feel wasteful, or cruel, or just clumsy.

> If we admire the truly wonderful power of scent by which the males of many insects find their females, can we admire the production for this single purpose of thousands of drones, which are utterly useless to the community for any other end, and which are ultimately slaughtered by their industrious and sterile sisters? It may be difficult, but we ought to admire the savage instinctive hatred of the queen-bee, which urges her instantly to destroy the young queens her daughters as soon as born, or to perish herself in the combat; for undoubtedly this is for the good of the community; and maternal love or maternal hatred, though the latter fortunately is most rare, is all the same to the inexorable principle of natural selection. If we admire the several ingenious contrivances, by which the flowers of the orchis and of many other plants are fertilised through insect agency, can we consider as equally perfect the elaboration by our fir-trees of dense clouds of pollen, in order that a few granules may be wafted by a chance breeze on to the ovules? (*Origin*, 202–3)

In the light of this aspect of the processes Darwin describes, enchantment cannot mean simply the sort of spiritual satisfaction that religion has attempted to provide, the ultimate assurance that behind the anguish and injustices of ordinary life there is a justice and a spiritual compensation that makes sense of that life at last, that rights the wrongs, that satisfies our sense of justice and our longings for peace. That organisms are diverse and often extravagant in form, that sheer and sublime mountains reflect billions of years of change, of earthquakes and eruptions and erosion, that there is a natural explanation for the useless protrusion from the fold in our ears and for the speciation of birds on the Galapagos may seem far from satisfying the yearnings of people who not only will never see or notice most of these phenomena but whose lives are daily and hourly consumed by the struggle for existence about which Darwin wrote. But they are part of everybody's life at the same time that they are obviously far larger than any individual life. They are awe-inspiring (as the natural history of the bee is awe-inspiring), but they are not always beautiful, and they do not always satisfy our longings for justice. Darwinian enchantment does not entail satisfaction of human desires; the sense of mystery that remains after we allow ourselves to believe that mysteries becomes puzzles and ultimately can be solved derives precisely from that powerful sense of difference that any close look at the natural world produces, and from the world's dogged resistance to allow itself to be subject to human wishes. As Joseph Vining has put it, while making a very different case from mine, the difference and uniqueness of the other—the not-me of the natural and human world—is a "presence" that "makes you not alone and who, in speaking, "is not yourself speaking to yourself." [5] This natural world promises no "justice"; it offers no meaning that is not in its own extrahuman processes. The fullest activity of human consciousness is required in even beginning to understand it, in recognizing that the world does not merely echo us, but speaks in a different language that we would do best to try to learn. And there is always more.

I have tried to demonstrate that while Darwin is given to an anthropomorphic vision, one that allows him to achieve a remarkable sympathy with the world while anchoring him still in the ordinary world of Victorian England, he is dramatically, almost testily, *not* anthropocentric. In reading his work as suffused with a sense of wonder, I have tried not to gloss over the fact that the world he represents to his readers and to the scientists who followed him was certainly subject to formulas of the kind that Dennett has used so successfully. The enchantment for which I take him as a model is dependent upon a recognition of those meanings, a recognition that the world is not "intelligently designed" for the ultimate good of us humans who inhabit it, that its workings proceed in absolute indifference to our desires and our intentions. He employed many of the argumentative strategies of Paley and the natural theologians; he achieved his deepest insights into the workings of nature in part because his imagination was shaped by those strategies. But he was entirely repelled by the egotism, the vanity, the self-complacency that insisted on understanding the phenomena of the world as somehow related to humans and to their betterment. Anthropocentrism was, he clearly thought, ultimately blinding. It doesn't allow us to see the part of the world that is not us, that is distinctive, different. And of course, the irony is that as Darwin looked around the world, freeing himself from that anthropocentrism, he found, miraculously enough, that even the most radical differences among organism imply—if you look closely enough, think hard enough—consanguinity.

Dennett claims to feel the sacredness of a world without a telos, and many scientists confronting the extraordinary complexity of natural laws governing the forms and development of the natural world probably share Dennett's sense of the sacred. The world "means" to those who attempt to understand, in the midst of the bewildering variety of natural events, even when they cannot yet explain some of them, or many of them. For those naturalistically inclined, the world may still feel like a mystery, but they address the mystery with the respect (and even affection) that is due to difference, while attempting to transform the mystery into a problem, confident that what *is* might eventually be explained naturalistically (but not worried that the explanation might disenchant the world, make it meaningless, or reduce the awe it inspires).[6] "Meaning" is what it is, how it is, where it is, in the ostensibly infinite complications of natural order, and "meaning," in another sense, lies in our respect for the peculiar natural condition that makes the thing what it is, and makes it distinctly not us. The conventional history of Western science narrates the gradual (or rapid) transformation of mysteries that had been the province of religion or of the supernatural in general into problems that became the province of science and natural explanation. (This kind of a narrative of progressive development gave the scientific naturalists of the last third of the nineteenth century an excuse for triumphalist rhetoric over the diminishing space left for the church and the supernatural.)

The excitement of their narrative, so frequently told with what was often felt to be arrogant aggression (we hear echoes of it in Pinker and Dennett, among others, in these latter days), is sufficient to account for scientists' enchantment while the nonscientific world tells itself that in the face of the scientists' explanation it is deadened and disenchanted. But there is the famous case of the suicide of George Price, a theoretical biologist who worked out a way "to measure the direction and speed of any selection process," and who discovered through that analysis, as he understood it, that whereas in any individual case altruism might be possible, "the human capacity for altruism must be strictly limited, and our capacity for cruelty, treachery and selfishness impossible to eradicate."[7] Finding a genetic basis for human idealism has been, through a large part of the twentieth century, something of an oxymoron. Only with the rebirth of the scientific belief in the possibility of group as well as individual selection has the idea of a genetically based altruism (other than "reciprocal altruism") been taken seriously by evolutionary theorists. Genetics had seemed to turn idealism into a fraud perpetrated simply so that the genes would increase the likelihood of their propagation. Of course, Price's was an extreme temperament, and his conversion to Christianity from atheism after his discovery emerged from a feeling of desperation. So it is not quite right to speculate that rigorously naturalistic and mathematical analysis drove him to his death. But it makes for a convincing parable. Reading human behavior back into the relentless and impersonal forces of natural law can be dispiriting and threatens always to reduce what feels like an ideal and a spiritual end to some version of a mere material drive.

The very language required to formulate the point, with its key verb "reduce," and the loaded word "threaten," and the implication that "material" is not only distinct from "spiritual" but morally inferior to it, almost necessitates disenchantment, or worse. But the rhetorical swerve here is not inevitable. Material meaning is after all "meaning," and there is no reason (except more than two thousand years of Western history, I suppose!) to decide that meaning is only meaning when it transcends the material conditions from which it emerges. Descartes was on to something important in Western thought, though he is often blamed for the mind/spirit dichotomy that he merely exploited. Certainly, his famous arguments about the separability of consciousness from its material home has done much to enforce the radically Christian view of the soul and body, and their comparative value, but the attitudes implicit in Descartes's rigorous reasoning have their affective counterpart in Price's (at least symbolic) death. The disenchantment narrative follows from the Cartesian dichotomy, and Price's suicide followed (at least logically) from the view that if altruism is simply a function of genetic inheritance it somehow loses its moral value.

Similarly, the introduction of chance into the Darwinian scheme was almost immediately taken (by nobody more than Darwin himself) as a radical intellectual and moral problem. Where a system is governed by chance, it is virtually impossible to decide on its "meaning," in the larger sense of the term, except the deeply disenchanted meaning that everything is chance, nothing has a reason. Darwin denied that chance played a role, though he could do nothing more than insist that it didn't. If chance mutation or chance variation is a condition of speciation and survival, the world suddenly becomes meaningless; and at the end of Darwin's century, George Bernard Shaw and Samuel Butler crusaded against the idea, although they were themselves radically secular and naturalistic thinkers. Butler brilliantly espoused a new Lamarckism, believing that Darwinism implied dead matter and an evolutionary dead end in mere repetition. For Butler all matter has at least unconscious memory, and out of that memory small, incremental, evolutionary changes proceed. "Matter which cannot remember is dead."[8] "I do not see," he says, "how action of any kind is conceivable without the supposition that every atom retains memory of certain antecedents." Such actions are the condition of evolution:

> We cannot believe that a system of self-reproducing associations should develop from the simplicity of the amoeba to the complexity of the human body without the presence of that memory which can alone account at once for the resemblances and the differences between successive generations, for the arising and the accumulation of divergences—for the tendency to differ and the tendency not to differ. (176)

If we were to talk about that "memory" today, we could do it through DNA and genetics, and in fact Butler was closer to the truth than the Lamarckian schema he opposed to Darwin would make it seem. But I turn to Butler here because I want to emphasize how critical it has been in Western self-understanding to discredit the Darwinian enterprise of describing the world as developing without the aid of consciousness and with no reference to intention and desire. (The past half-century's work in DNA and genetics has remarkably accounted for the "memory" Butler rightly sought, without needing the consciousness and intention that Butler assumed would be necessary.) I focus on Butler in particular because he was, on the one hand, entirely committed to the idea of evolution and the gradualist process that Darwin had described and, on the other, violently opposed to Darwin on the grounds that his explanation of *how* this happens emptied the world of meaning—for him, the secularist, as much as for religious believers who resisted any sort of naturalistic explanation of the human. Butler's appalled resistance to Darwin's theory suggests that the need for some sense of "meaning" beyond the merely descriptive one of scientific explanation extends well beyond the world of religion.

The crisis of "meaning" as Western thought has traditionally formulated it requires some inclusion of intention and design in the processes of nature, even if the traditional intention and design of natural theology are entirely rejected. We have seen that even Darwin introduced intention into his theory of sexual selection (in ways, in fact, at least partly compatible with Butler's). But Butler responds to Darwin's theory with the quite reasonable demands of a much wider (and less scientific) public: how can the world matter if it is not anthropocentric, has no relation to human intention and desire, and no teleology? Butler, then, resists the work of disenchantment by infusing the *natural world* with "intention."

In the course of thinking about this book and my arguments for "enchantment" within the frame of Darwinian (not Butlerian) science, I have tried to understand what it is that Butler wants when he seeks meaning in the evolutionary world, and what it is that people mean when they ask, what does it all mean? For Butler meaning inheres in a very human answer: the world means insofar as it works with a more or less human consciousness guiding it. (This attitude seems—though it is for Butler strictly naturalistic—very similar to the attitude that drove and drives natural theology, and now, the anti-Darwinian theory of "intelligent design." Natural theology can only explain the evidences of design—Paley's "watch" found among stones—as the product of a humanlike intention; theorists of "intelligent design" refuse to accept the possibility that complex structures and processes in the natural world can be explained without the invocation of "intelligence," a quality distinctly human.[9]) While he goes on to describe consciousness as very different from the human, inhering in chickens as well as in atoms, the world newly invested with intention enacts, on the whole, human desires— the desire, in particular, to believe that things don't just happen but develop in response to consciousness, desire, and need.

Butler's world, then, is radically anthropocentric. As against the criticism that Darwin's arguments are too often anthropomorphic, it is striking to realize that virtually all critiques of his work from the side of culture and religion, and many from the side of science itself, have been driven by a powerfully *anthropocentric* interest against which Darwin's "anthropomorphism" was designed to work.

The Butlerian view reintroduces responsibility into the system not by appealing beyond nature but by insisting that the system moves only because the organisms (including human organisms) who populate it consciously or (for the most part) unconsciously move it. Humans evolve as they drive themselves toward increasingly perfect ends. Butler reintroduces teleology into the system from which Darwin had expelled it, though he does it without divine assistance.

What kind of answer would satisfy most who ask the question of what it all means? A theodicy, or, more properly an anthropodicy (ouch). A theodicy, as we have seen, provides an explanation of all the evil and pains of the world as part of a greater plan of ultimately divine justice: everything will, under the aspect of eternity, at least, work out okay. Evil is a test of our spirits, perhaps, or in any case can be explained in terms that satisfy our human sense of justice. The justification of God's ways to man is built on an explanation that ultimately satisfies man, provides the justice and fairness that is so often missing from our natural lives. Naturalism can't provide such an explanation; things work out regardless of human desire, entirely as the laws of nature allow them to work out. Darwin is himself uneasy with this hard truth and awkwardly tells us that at least we may take comfort that the death of organisms is generally "swift," and that they do not suffer too much pain. But in the virtually eternal debates that swirl around such questions, we must return again to what may have been a defining event in Darwin's life: nothing explains away the painful death of Annie Darwin. Was it a retribution for some past sins of hers? Was it a test of the family's faith? Was it a retribution for the sins of her heretic father? (Darwin had, after all, written but not published drafts of his theory by the time Annie died.) Even to someone with merely human values, that sort of retribution must seem morally worse than the dead indifference of the disenchanted world the culture laments. Personally, I would rather have the simple Darwinian meaning: Annie died from sheer random movements of a virus, or from debility resulting from the genetic similarity of her parents, or from other natural causes. Deference to a power we cannot understand in the faith that it all makes anthropocentric sense seems a very weak option indeed.

Obviously, without some such teleology, the joys that a detailed understanding of nature might bring, the sheer thrill of sublime engagement with natural phenomena, can do nothing to ease the pain of the loss of a daughter. But putting aside the attempts to justify God's ways to man, or the attempts to show that in the very long run humans are responsible for a progressive move toward some higher condition, where the ape and the tiger die, Darwinian secularity simply acknowledges the workings of natural law and offers us an earth that must be room enough. Refusing the work of justification and the imagination of teleology, Darwin's theory explains as well as possible how the material world works within a natural order, without particular reference to human need or intention, and in so doing the theory also demonstrates the absolutely remarkable conditions that govern life in this world. Those penguins have figured it out for themselves: what does it take to breed safely in this awful cold? How do I recognize my mate's voice in a hubbub of hungry penguins? These penguins are not very well designed, nor are they anthropocentric parables of monogamy or redemption. They are penguins.

Perhaps there is some compensation in the recognition that the very values that prod us so often to look beyond nature are the products of nature. When Darwin notes the surprising ways in which organisms adapted to one kind of life perform successfully in another, something of the wonderful strangeness of life has to register with every reader. To know that men's nipples are vestiges of a hermaphroditic past puts men, I would think, in a strangely remystified relation to their bodies.

But the question remains: what sort of answer, if not these "natural" ones, would satisfy the question, What's the point? What does it mean? As we speak of the disenchantment that follows upon the scientific explanatory mode, we mean surely that we have lost the answer that would identify all events in the heavens and the earth, the fall of every hair from the head, the acorns dropping in otherwise silent forests, the genocides that punctuate history, as somehow—however indirectly, and with whatever losses—pointing toward the satisfactions of our very human sense of justice and fairness, our very human sense that we want in the end to find joy. So the resolution of the question of what the world means would depend on the answer's power to demonstrate that human needs are being satisfied.

Darwin's way of thinking suggests that the question has to be changed, or the meaning of "meaning" reunderstood. The most exciting thing about Darwin's theory, perhaps, is that it is *not* anthropocentric. It risks *not* making the human the be-all and end-all of nature, but it is at the same time overwhelmingly human, down to the inevitable contingency of every human condition. If Darwin is to be blamed for being anthropomorphic, we might (as he implicitly did in his personal critique of natural theology) criticize the religious view of the world or even Butler's secular version of it for being almost outrageously anthropocentric. Refusing to privilege the human or exempt humans from the workings of nature, Darwin's work points toward the possibility of a nonhierarchical imagination of the human condition. (As throughout this book, I want again to separate out the possibilities intimated and asserted by Darwin's remarkable work from the very particular, Victorian, hierarchical biases that are evident in it. I have tried to make clear that he was always the Victorian gentleman, with the Victorian gentleman's cultural values, but that his implication in his particular historical moment not only did not damage his science but, in many respects, was a condition for it.)

It points toward a new meaning for "meaning." Refusing anthropocentrism, even as he continued to think anthropomorphically, Darwin suggests a liberating condition in which one's inevitable implication in one's culture is not incompatible with an imagination and intelligence that can point to ways to break free from it. Darwin's world "means" itself; it means its complicated, diverse connectedness. And Darwin provides us with the intellectual tools that will allow us to come to terms with it.

We can't help seeing with the eyes that are contingently located at this moment in this place in this body. We almost can't help imposing on the world—Feuerbach would have said "projecting on" it—our own values and our own images. The problem, Feuerbach claimed and, I would say, Darwin enacted, is that we have too often come to worship our own projection while assuming or pretending that we are worshiping an Other. Darwin anthropomorphized, in part as a way to allow himself, and us, to feel at home in a world without "Meaning" with a capital M. He could love, but he would not worship or falsify, what he projected.

To move from a notion of a disenchanted world to a new experience of enchantment one needs to begin by rejecting anthropocentrism and coming to terms with the fact of nonhuman diversity, by recognizing perhaps in a Feuerbachian way that the values that inhere in the condition of prescientific enchantment were not only theological but powerfully anthropocentric. Insofar as we want to privilege those values, we need to recognize that it is human consciousness and experience itself that is putting them there, that transforms mere matter into the sublime and the beautiful. And we need to recognize that those values, emerging from our own desires and needs, perhaps even from our own development through natural selection, are no less valuable for being absolutely human.

Does the sunrise become less beautiful because scientists can explain its movements (of all sorts) and the way it produces its colors? In fact, John Stuart Mill, an early victim of disenchantment, it seems, was forced to address that problem. Responding to a utilitarian friend for whom, it seems, all knowledge was instrumental, Mill writes:

> It was in vain I urged on him that the imaginative emotion which an idea when vividly conceived excites in us, is not an illusion but a fact, as real as any of the other qualities of objects, and far from implying anything erroneous and delusive in our mental apprehension of the object, is quite consistent with the most accurate knowledge and most perfect practical recognition of all its physical and intellectual laws and relations. The intensest feeling of the beauty of a cloud lighted by the setting sun, is no hindrance to my knowing that the cloud is vapour of water, subject to all the laws of vapours in a state of suspension; and I am just as likely to allow for, and act on, these laws whenever there is occasion to do so, as if I had been incapable of perceiving any distinction between beauty and ugliness.[10]

Mill wants here to combine the rational and the practical with the sense of "beauty and ugliness," of awe and wonder, that he and most other people are likely to experience in the presence of a splendid sunset. For Mill, emerged from that famous depression that has something about it of the feeling of disenchantment, knowledge and scientific analysis are not incompatible with a sense of noninstrumental value. Do humans become less important because we now know they have emerged over vast sweeps of time from nonhuman ancestors? Darwin claimed, as we have seen, that the recognition made him feel "ennobled." What a "miracle" that time (and, yes, chance, as Stephen Jay Gould constantly reminded us) and the laws of nature might produce from a hairy quadruped, and ultimately from a hermaphroditic monocellular organism, such a stunningly complex and powerful and vulnerable being.

But still, none of this compensates for the loss of a daughter. The anthropodicy assumes such losses, in a way explains them without justifying them morally, frees us from a need to make things mean within a moral scheme that does not really allow for the otherness of the world, for its stunning and moving resistance to our assumptions and prejudices. Just because Annie's death was not justified according to the values we take to be distinctively human, we can free ourselves from the constraining need to *make* it just. None of this resolves the awful questions or improves the awful conditions and inequities that mark so much of human life. But it is a beginning toward a way of coming to terms with a world so much larger than any individual and so indifferent, except as we ourselves choose not to be, to our individual fates. This world we never made but which, for better or worse, we are always in the process of remaking needs to be confronted with a sense that even without "intention," it is wonderful, more wonderful just because it is what it is without an intention like our own working upon it. The physiologist W. B. Carpenter wrote in 1858 about the miracles he could discern in the minutiae under the microscope, and his vision there, as the invisible world opened up under the new technology, suggests a Darwinian sublime: "There is something in the extreme of minuteness, which is no less wonderful—might it not almost be said, no less majestic?—than the extreme of vastness."[11] Darwin found that to be true of barnacles, and ants, and worms and bees, so that for him, looking with intensity and patience at the minutest of creatures, the ordinary was sublime. Darwin transferred reverence to the ordinary, and it is partly that work that makes him so excellent a model for secular enchantment. With those remarkable powers of observation, he saw the world in a grain of sand, and he saw diversity where those who do not look closely, those who look beyond the earth, see only sameness. One of the world's primary values in the Darwinian vision, one of the very conditions of our being, is the stunning diversity that he described for us in painstaking and often beautiful detail. Confronting the not-ourselves everywhere, we find, as Darwin taught us, that we are both a part of what we do not recognize— remember Darwin's shock at the sight of those Fuegians, from whom he recoiled because he knew instantly that they were fellow humans—and wonderfully different as well.

Darwinian enchantment entails awe and love of that diversity, the recognition that upon it depends our own sharp distinctiveness. As we have seen, Darwin's shock and horror at the first Fuegian encounter never disappeared. But if he could never come to love the Fuegians or to believe that they were evolutionarily the equal of his own race and class, he created a "science" that demonstrated kinship—not spiritual, but literal blood relationship—I prefer nowadays "genetic" relationship. He opened, then, also a possibility of love, revealing that nature might produce a diversity that disguised what many anthropocentrics didn't want to know. "Am I not a man and a brother?"

I began with a bumper sticker, facetious and ironic, perhaps, but also serious. "Darwin Loves You" is a claim for an enchanted secularity. I have taken the risk of naming this book after that bumper sticker, not because I think it's funny but because reading Darwin has been for me a kind of secular epiphany. (The resort to religious language is inescapable, and part of the point. The world Darwin reveals is one that is worthy of the awe, wonder, admiration, and fear that has traditionally been arrogated to the world of religion.) I realized from the first that with all his quite touching efforts to remain the dispassionate observer, he never fully managed in the *Origin* to divorce himself from the materials he represented. It is evident everywhere that Darwin had a feeling for the organism, a remarkable capacity to think himself into the creatures whose history he was trying to understand. The powers of observation that he humbly allows he had were partly conditioned by his passion for his subject. Yes, one must reiterate, some of the fundamental Victorian biases and assumptions, of his class, of his gender, of his race, went into the creation of his "long argument," and it is not unreasonable, as many current historians of science think, to believe that those aspects of his thought are not merely contingent, as I have been maintaining, but, as Janet Browne has claimed, "constitutive." But contingent or constitutive, they did not impede Darwin from the development of his theory, which through a now long history of ups and downs, has proven to be extraordinarily fruitful. Even if we accept the notion that his cultural assumptions are constitutive, then we must accept the fact that Darwin's enchanted vision of a mindless and chance-driven world is not only possible but intrinsic, that the feelings of awe and wonder are fully compatible with scientific clarity and rational precision. I have filtered out of Darwin's writing and life a kinder, gentler Darwin. But it matters more that this filtered Darwin provide a usable model of nontheistic enchantment than whether the filtering does full justice to the complexity and culture-bound nature of his thought and life.

Darwin is no perfect model, but the absence of perfection, the fact that he is no saint but distinctly in many respects an ordinary (if wealthy and talented) man makes him an almost perfect figure with which to focus the possibilities of secular enchantment. Just as his work concentrates heavily on the most ordinary organisms and makes us understand them in new and enchanting ways, so I concentrate in this book on an ordinary man whose work managed to transform late-century science and whose loving and imaginative engagement with the natural world can provide the necessary evidence that enchantment and scientific rationalism can coexist. We have seen that Darwin's prose was often enchanted, registering the wonder and awe that Darwin felt in his explorations of the world. He found a language to represent it that depended in part on leaps of imaginative sympathy and on a humility that allowed him, sometimes against his own strong cultural prejudices, to listen to it, to feel for it, to see it clearly. He had a feeling for the organism. Despite the weariness of his later years, despite his illness, despite his losses, Darwin very clearly loved the world and loved its differences. Darwinian enchantment entails a direct and equal confrontation with the myriad otherness that constitutes the evolved world. Out of barnacles, sea-slugs, ants, worms and vegetable mould, and climbing plants, he created a sublime of the ordinary, or rather, as he himself perhaps would have preferred to put it, he *encountered* a sublime of the ordinary that evoked awe and wonder and a sense of mystery.

This Darwin is the one featured on that bumper sticker my son gave me. It is the Darwin who does not empty the world of spirit but fills it with wonder. Certainly, it is only a small step that Darwin, as a model of nontheistic enchantment, can help us take. But in a world living inside the myth of disenchantment, producing false gods (or sky hooks) in reaction, it is critical that the possibility of secular enchantment be affirmed—and experienced. Seeking that kind of feeling that is essential to ethical action, secularists will do well to read Darwin with something like the critical attention he gave to his barnacles and worms. I have read him here not only with an eye toward his own cultural prejudices, of which there were multitudes, but as a kind of first exemplar of a new possibility of joy in this world, and love of it—even with the sternest and most intelligent recognition of its "algorithmic" functioning. An enchanted, secular vision seeks a feeling for the organism, values the extraordinary differences that mark the range of organic life, and depends on imagination (anthropomorphic perhaps, but not anthropocentric), honoring difference, recognizing penguins for penguins. In a world so overwhelmed as ours by catastrophes and inhumanity, enchantment may seem a trivial thing. But enchantment is a condition of living at home in this world we never made (but are making day by day), and loving it, just as the bumper sticker my son gave me announces that Darwin loves *you*.

NOTES

PREFACE

1. In an interesting and impressive essay on the subject, Gregory Radick argues for what he calls the "inseparability thesis," the thesis that the contingent conditions that drive the development of an argument are inseparable from that theory. So the Malthusian connection to Darwin's theory of natural selection is intrinsic to it. Logically, the argument is very convincing, and in certain particular senses it has to be right. On the other hand, it is simply historical fact that many thinkers have in fact dissociated natural selection from Malthus. The undoubtedly correct reading of this is that in some important ways, these uses of natural selection are not truly Darwinian. But my point is that many of these thinkers claimed to *be* Darwinian nevertheless. Moreover, there is another problem. Radick rightly points out that as modern evolutionary theorists now understand the theory, "selection occurs whether or not resources are scarce. All that matters is that there are differences of fitness within a population." The question for this book is whether the new understanding of natural selection could be called Darwinian. I would argue here—and much of this book is based on this view—that in any useful sense, it remains Darwinian, even if modified. It accepts the idea of natural selection and simply replaces the notion of Malthusian "struggle" with the notion of fitness differential, which remains based on a Malthusian model, though the emphasis on direct struggle is diminished. So while I understand and credit the argument of the constitutive nature of those contingent forces I emphasize, I insist that history changes the terms and complicates the issue. Much that may not be constitutive Darwinism has historically made its claim to be Darwinian. See "Is the Theory of Natural Selection Independent of Its History?" in *The Cambridge Companion to Darwin*, ed. Jonathan Hodge and Gregory Radick (Cambridge: Cambridge University Press, 2003), 143–67. See esp., 157–59.

2. Charles Taylor, "Modes of Secularism," in *Secularism and Its Critics*, ed., Rajeev Bhargava (New Delhi: Oxford University Press, 1998), 53. In this important essay Taylor distinguishes several modes of secularism; he is concerned to work out ways in which fundamental and divergent religious beliefs may function within a politically coherent and peaceful polity. He argues that only a secularism that functions without the need of some fundamental "commonly held foundation" can any longer be expected to work. His conclusion is that to survive, secularism must function according to what he calls "overlapping consensus"; such a secularism, he asserts, and I entirely agree, is absolutely essential to the survival of modern democratic states: "either the civilized coexistence of diverse groups, or new forms of savagery. It is in this sense that secularism is not optional in the modern age" (48).

CHAPTER 1

1. William Hurrell Mallock, *Is Life Worth Living?* (London: Chatto and Windus, 1879), 17.

2. Stephen Pinker, *How the Mind Works* (New York: W. W. Norton, 1997), ix.

3. Max Weber, "Science as a Vocation," in *Max Weber: Essays in Sociology*, trans. and ed. H. H. Gerth and C. Wright Mills (New York: Oxford University Press, 1946), 139.

4. William James, *The Varieties of Religious Experience* (1902; New York: Modern Library, 1994), 12.

5. Pinker is perhaps the most popular and yet the most unrelentingly disenchanting of current American popularizers of evolutionary psychology. On pages 306–407 of *How the Mind Works* he neatly summarizes the views in evolutionary psychology that read altruism into Darwin's theory of natural selection. He fully accepts the dominant view that group selection virtually never operates in nature although it is "possible on paper" (397). To understand something of the triumphal rationalist rhetoric that marks the disenchanting view of the world, I quote here Pinker's definition of "love": "When an animal behaves to benefit another animal at a cost to itself biologists call it altruism. When altruism evolves because the altruist is related to the beneficiary so the altruism-causing gene benefits itself, they call it gene selection. But when we look into the psychology of the animal doing the behaving, we can give the phenomenon another name: love" (400).

6. When James defines "religion" as *"feelings, acts, and experiences of individual men in their solitude, so far as they apprehend themselves to stand in relation to whatever they may consider the divine,"* he leaves open (and quite intentionally) the possibility that religious experience might even be "atheistic," as, for example, he describes both Buddhism and Emerson's quite religious sense of life. "We must therefore, from the experiential point of view, call these godless or quasi-godless creeds religion; . . . we must interpret the term 'divine' very broadly, as denoting any object that is god*like*, whether it be a concrete deity or not" (39–40).

7. George Eliot, *Middlemarch: A Study of Provincial Life* (1872; Oxford: Oxford University Press, 1996), part 1, chapter 3, 24–25.

8. James Moore has shown how important it was to Darwin's supporters and contemporaries to remove political implications from the theory. The language of "Darwinism" was laden with implications that, to ensure the full scientific acceptance of evolution, T. H. Huxley tried to avoid. Moore shows how Huxley achieved successes in part by convincing people that Darwinism was "neutral" on questions of "sociology and politics." James Moore, "Deconstructing Darwinism: The Politics of Evolution in the 1860s." *Journal of the History of Biology* 24/3 (Fall 1991): 353–408.

9. John Durant, introduction to *Darwinism and Divinity: Essays on Evolution and Religious Belief*, ed. John Durant (Oxford: Basil Blackwell, 1985), 2.

10. Mary Midgley, "The Religion of Evolution," in Durant, 154.

11. Charles Darwin, *On the Origin of Species by Means of Natural Selection, or the Preservation of Favoured Races in the Struggle for Life*, a facsimile of the first edition (1859; Cambridge, MA: Harvard University Press, 1964), 79.

12. Diane B. Paul, "Darwin, Social Darwinism and Eugenics," in Hodge and Radick, 214–15.

13. Stephen Shapin and Barry Barnes, "Darwin and Social Darwinism: Purity and History," in Barry Barnes and Steven Shapin, *Natural Order* (Beverly Hills: Sage Publications, 1979), 125–39.

14. *The Correspondence of Charles Darwin*, ed. Frederick Burkhardt et al. (Cambridge: Cambridge University Press, 1985–), 2:444.

15. Aside from the essay by Robert Young, cited below, there are many excellent essays and books making the case. Among others, see John C. Greene, *Science, Ideology and World View: Essays in the History of Evolutionary Ideas* (Berkeley: University of California Press, 1981); Sylvan Schweber, "Darwin and the Political Economist: Divergence of Character," *Journal of the History of Biology* 13 (1980): 195–289.

16. For a full discussion and for references, see my *Dying to Know: Narrative and Scientific Epistemology in Victorian England* (Chicago: University of Chicago Press, 2002).

17. Adrian Desmond and James Moore, *Charles Darwin: The Life of a Tormented Evolutionist* (New York: Warner Books, 2001), 267.

18. Adrian Desmond and James Moore, eds., *The Descent of Man*, by Charles Darwin, 2d ed. (1874; London: Penguin Books, 2004), lvi.

19. Matt Ridley, *The Origin of Virtue: Human Instincts and the Evolution of Cooperation* (London: Penguin Books, 1996), 260.

20. It might be unfair at this point to invoke a later book by another and distinguished scientific author for "counterevidence" on the question of human interference in the processes of nature. But it is useful to notice what different sorts of ideological morals might be educed from the "facts" of nature and rigorously "Darwinian" thinking. Jared Diamond's *Collapse* (New York: Viking, 2005) makes just the reverse political case from Ridley's, condemning a pure ideology of laissez-faire as he demonstrates how twelve civilizations collapsed by virtue of following out their "natural" likes and dislikes, their "natural" desires for certain kinds of material objects.

21. Throughout this book I will have many occasions to allude to these two major biographies: Desmond and Moore, *Charles Darwin* (see note 17), and Janet Browne's two-volume study, *Charles Darwin: Voyaging* (New York: Knopf, 1985) and *Charles Darwin: The Power of Place* (New York: Knopf, 2002).

22. Adrian Desmond, *The Politics of Evolution: Morphology, Medicine, and Reform in Radical London* (Chicago: University of Chicago Press, 1989), 412.

23. James Secord, *Victorian Sensation: The Extraordinary Publication, Reception, and Secret Authorship of* Vestiges of the Natural History of Creation (Chicago: University of Chicago Press, 2000).

24. Loren Eiseley, *Darwin's Century: Evolution and the Men Who Discovered It* (1958; New York: Doubleday Anchor, 1961), 196.

25. See Dov Ospovat, *The Development of Darwin's Theory: Natural History, Natural Theology and Natural Selection, 1838–1859* (Cambridge: Cambridge University Press, 1981). Ospovat shows that "for many years, and in some respects throughout his life, Darwin shared his contemporaries' belief in harmony and perfection" (3), but that between 1844 and 1859 Darwin began breaking with this natural-theological view of adaptation, until by the time the theory was fully formulated, "the idea of perfect adaptation played no role" (4). Nevertheless, the shape of the theory was largely determined by the questions a natural-theological view posed, and the centrality of the question of the "production of adaptation" is distinctly informed by natural theology.

26. James Secord, "Response," *Journal of Victorian Culture* 8 (Spring 2003): 148.

27. Derek Attridge, *The Singularity of Literature* (London: Routledge, 2004), 22.

28. For a compact and careful analysis of the differences between Dawkins and Gould, see Kim Sterelny, *Dawkins vs. Gould: Survival of the Fittest* (Cambridge: Icon Books, 2001). Sterelny begins by pointing out a very broad range of agreement between the two, all of the points being distinctly Darwinian. The differences derive from fundamentally diverse emphases in the reading and uses of Darwin but have, as Sterelny demonstrates, important implications for large cultural questions.

29. In a subtle and careful historical reconstruction, Robert Richards rejects the modern consensus that Darwin's theory, even in the *Origin*, was not progressivist in some ways. Richard demonstrates that implicit in Darwin's theory was a sense of evolution as connected with embryology and with the famous idea, which Darwin never specifically affirmed, that ontogeny recapitulates phylogeny—a fundamentally progressivist idea largely connected to Haeckel. From his earliest notebooks, as Jim Endersby shows in "Darwin on Generation, Pangenesis and Sexual Selection," "Darwin accepted that individual maturation repeated, and recapitulated . . . the changes in form gone through by the species' entire ancestry since life began" (Hodge and Radick, 73). To be fair, then, if Richards is correct about Darwin and "evolution" as connected with embryology, we have another case in which readers with particular historical and ideological engagements create the "great" text. See Robert Richards, *The Meaning of Evolution: The Morphological Construction and Ideological Reconstruction of Darwin's Theory* (Chicago: University of Chicago Press, 1992). See in particular chapter 6 for a discussion of the way interpretation is inflected by "ideology."

30. Perhaps the formulation most significant to the modern synthesis is Ronald Fisher, *The Genetical Theory of Natural Selection* (1930).

31. For a survey of many of the various ways the interpretation and uses of Darwin's ideas were shaped by place and social context, see Ronald L. Numbers and John Stenhouse, eds., *Disseminating Darwinism: The Role of Place, Race, Religion, and Gender* (Cambridge: Cambridge University Press, 1999). In addition, Ospovat insists that Darwin shared with his contemporaries a belief in "progress." While he clearly did not believe that for any given individual or species progress was inevitable, he did believe that "natural selection does not exclude the inevitability of various general results." Ospovat quotes Darwin from *The Descent of Man*: "any animal whatever, endowed with well-marked social instinct, the parental and filial affections being here included, would inevitably acquire a moral sense or conscience, as soon as its intellectual powers had become as well, or nearly as well developed, as in man" (225).

32. One of the most powerful and convincing elaborations of this argument is Secord's. Secord is concerned with the context of Robert Chambers's text, not only as it was written but also—and primarily—as it was read, for, as he claims, "The texts of science have no meaning apart from what readers make out of them" (532). To understand what they made of them entails an almost endless proliferation of readings of readings of readings, which themselves entail an attempt to reconstruct (out of readings, of course) the contexts that helped determine the shape of the readings. While there is much to be debated

in this view of "texts," Secord's brilliant and meticulous historical reconstructions can throw light too on Darwin's "text," which has undergone and continues to undergo such contextual recreations and reinterpretations, of which, of course, the ones discussed in this book are only a small part. It is important to note that in his "Response" to critiques of his book, Secord is careful to argue that "no text should ever be reduced to its context." His sense of the diversity of "culture" disallows the possibility of a single direction that cultural context might determine.

33. Robert M. Young, "Darwinism *Is* Social," in *The Darwinian Heritage*, ed. David Kohn (Princeton: Princeton University Press, 1985), 609–40. In exasperation, Young complains that it is still necessary to argue that "The intellectual origins of the theory of evolution by natural selection are inseparable from social, economic and ideological issues in nineteenth-century Britain" (609).

34. This is not the place nor the book in which to enter the debate on the philosophical implications of such historicism. The radical implication of Secord's view that "the texts of science have no meaning apart from what readers make out of them" would seem to move toward a Rortyan pragmatist view of meaning. For Rorty, of course, language does not represent (the way nature really is); its value is determined by whether it works. It is "right" or "wrong," "good" or "bad," not because it represents reality but because in the particular historical conditions in which it is uttered it turns out to be more or less useful for the specific purposes of the speakers and their auditors. A Rortyish reading of texts like Chambers's or Darwin's, then, would, I presume, focus on those historical conditions and see the language as having meaning only within the conversations in which they take part. For the purposes of this book, the practical implications of such epistemology (or anti-epistemology) are not important, though the general issues of meaning are, of course, quite serious. What is most useful for my argument is the way in which such theories return us to both historical contingency and human usefulness. For one of many statements by Rorty on the question of scientific language, see *The Consequences of Pragmatism* (Minneapolis: University of Minnesota Press, 1982), particularly the essay "Method, Social Science and Hope," 191–210.

35. Stephen Jay Gould, "More Things in Heaven and Earth," in *Alas, Poor Darwin*, ed. Hilary and Steven Rose (New York: Harmony Books, 2000), 104.

36. See William Connolly, *Why I Am Not a Secularist* (Minneapolis: University of Minnesota Press, 1999), 16.

37. Jane Bennett, *The Enchantment of Modern Life: Attachments, Crossings and Ethics* (Princeton: Princeton University Press, 2001), 10.

38. John Herschel, *A Preliminary Discourse on the Study of Natural Philosophy* (1830; Chicago: University of Chicago Press, 1987), 15.

39. Charles Taylor, *Philosophy and the Human Sciences*, vol. 2 of *Philosophical Papers* (Cambridge: Cambridge University Press, 1985), 276.

40. John Stuart Mill, "Nature," in *"Nature" and "Utility of Religion"* (1874; Indianapolis: Bobbs-Merrill, 1958), 20.

41. Taylor's brief summary of the movement from an enchanted world in which the sacred inhered in nature and in the regular religious rituals that marked the movements of ordinary life to a world from which the sacred was banished is helpful here. Sacred places and times are those "in which the divine

or the holy is present." In an enchanted world of this kind, the sacred might be thought to inhere in the polity itself, and in particular in its leader, its king. But with the advancement of "disenchantment, especially in Protestant societies, another model took shape.... In this, the notion of design was crucial." Instead of the sacred inhering *in* the world, the world was "conceived in conformity with post-Newtonian science, in which there is absolutely no possibility of higher meanings being *expressed* in the universe." But the universe gives evidence of God's presence in its design. Charles Taylor, *Varieties of Religion Today* (Cambridge, MA: Harvard University Press, 2002), 64, 66. The movement from the sacred, through Protestantism and science, to secularism is partly completed by Darwin's theory. The last move of Darwin, of course, is to deny that "design" gives evidence of God's presence; but, I would argue, the move from sacred expression to evidence of God's presence to the absence of intention from design has an *affective* continuity. Darwinian enchantment is built on this continuity and provides the special emotional and ethical satisfaction of feeling-saturated thought.

42. Hilary Putnam, *The Collapse of the Fact/Value Dichotomy and Other Essays* (Cambridge, MA: Harvard University Press, 2002), 26.

43. It is, or should be, impossible to consider Darwin in this way without attention to the work of Gillian Beer, who, with a rich understanding of the substantive arguments Darwin makes, attends to his rhetoric and to the cultural resources with which he works. See Gillian Beer, *Darwin's Plots: Evolutionary Narrative in Darwin, George Eliot and Nineteenth-Century Fiction*, 2d ed. (Cambridge: Cambridge University Press, 2000); "Darwin's Reading and the Fictions of Development," in Kohn, 543–88.

44. Quoted in James Moore, "Darwin of Down: The Evolutionist as Squarson-Naturalist," in Kohn, 407.

45. Peter Allan Dale, *In Pursuit of a Scientific Culture* (Madison: University of Wisconsin Press, 1989), 4–6.

46. T. R. Wright, *The Religion of Humanity* (Cambridge: Cambridge University Press, 1986), 4.

47. Max Weber, "Science as a Vocation," in Gerth and Mills, 142.

48. William James, "The Will to Believe," in *The Will to Believe and Other Essays in Popular Philosophy* (1897; New York: Dover Publications, 1956), 9.

49. I have always felt more than a sneaking affection for Clifford just because he is so passionate about rationality. This is not a paradox but in fact consistent with both James's and Putnam's sense of the nature of epistemological work. Clifford believes, with almost religious intensity, in the moral necessity of epistemological austerity, and the problem for his reputation and for that of his contemporary positivists is that their "passion" for rationality was not something that any but a small cluster of intellectuals could share. As in his lecture "The Ethics of Belief," Clifford puts epistemological questions to moral tests: should, for example, a ship owner, without sufficient evidence of its seaworthiness, allow his merchant craft to go to sea? If he does, people may die because of the decision. And this is true even if the ship owner truly believed that the ship was seaworthy. "No man," Clifford says, "holding a strong belief on one side of a question, or even wishing to hold a belief on one side, can investigate it with such fairness and completeness as if he were really in doubt

and unbiassed; so that the existence of a belief not founded on fair inqury un-fits a man for the performance of this necessary duty." *Lectures and Essays*, (London: Macmillan and Co., 1902), 2:168.

50. See also 34, 45–46. "Proponents of disenchantment share with those who lament its loss the assumption that only a teleological world is worthy of our enchantment" (11).

51. In Richard Dawkins, *The Selfish Gene* (1976; Oxford: Oxford University Press, 1999), a book that is often taken as the central text in the contemporary arguments for the primacy of natural selection in the explanation of life, in-cluding human life, Dawkins specifically refuses to engage the concept of al-truism in the psychological or moral sense. For him, to be altruistic is to behave "in such a way as to increase another such entity's welfare at the expense of [one's] own" (4). The word "selfish" in the title is a kind of provocation, but Dawkins is really not interested in the problem of whether some people be-have with genuine altruistic intentions. He is concerned only with the "real ef-fects" of behavior on the survival of the species. "Altruism" is the problem that *The Selfish Gene* was intended to resolve; debates within the community com-mitted to the dominance of natural selection in evolution have pushed increas-ingly toward the possibility of real altruism as arguments for the possibility of group selection have slowly begun to become respectable. See, in particular, David Sloan Wilson, *Darwin's Cathedral* (Chicago: University of Chicago Press, 2002): Wilson is totally committed to naturalistic explanations of altruistic be-havior but finds that altruism is, indeed, more productive in the success of groups than is selfishness. And against the dominant current of evolutionary biology, Wilson argues impressively for the possibility of group selection.

52. Charles Darwin, *The Autobiography of Charles Darwin, 1809–1882*, ed. Nora Barlow (1887; New York: W. W. Norton, 1958), 90.

53. See Robert J. Richards, *The Romantic Conception of Life: Science and Philoso-phy in the Age of Goethe* (Chicago: University of Chicago Press, 2004).

54. Charles Darwin, *The Descent of Man, and Selection in Relation to Sex* (Princeton: Princeton University Press, 1981), 17.

CHAPTER 2

1. Robert M. Young, "Darwin and the Genre of Biography," in *One Culture: Essays in Science and Literature*, ed., George Levine (Madison: University of Wis-consin Press, 1987), 210.

2. I don't in this book want to get entangled in the now long-standing de-bates about "intrinsic" and "extrinsic" explanation of science. I need at every point to reemphasize that while I am committed to the idea that there are ways of talking about how science works (more or less effectively, producing better or worse information), I share with Young, Desmond, and Moore the view that science is "enmeshed" in culture. In some sense, it is always inadequate to in-voke the science *or* culture distinction, but there is no simple way to avoid talk-ing about the science, and then about the cultural influences on the science. This would be harmless enough if it were understood that neither exists in a pure form. Pierre Bordieu seems to me correct when he says: "An authentic sci-ence of science cannot be constituted unless it radically challenges the abstract

opposition . . . between immanent or internal analysis, regarded as the province of the epistemologist, which recreates the logic by which science creates its specific problems, and external analysis, which relates those problems to the social conditions of their appearance." Pierre Bordieu, "The Specificity of the Scientific Field and the Social Conditions of the Progress of Reason," reprinted in *The Science Studies Reader*, ed. Mario Biagioli (New York: Routledge, 1999), 33.

3. Alasdair MacIntyre has claimed that the is/ought dichotomy, at the heart of the Enlightenment project, does not hold, that there are many facts about "functional concepts"—that is, words that describe agents who are understood in relation to what they do, like "farmer," for example—that entail value judgments. And, MacIntyre argues, "man" is itself a functional concept: man "understood as having an essential nature and an essential purpose or function." Only, MacIntyre claims, when this notion is rejected, when "good man" gets separated from "man," does the is/ought dichotomy come into play. See *The MacIntyre Reader*, ed. Kelvin Knight (South Bend, IN: University of Notre Dame Press, 1998), 79–80.

4. In a still important and interesting study of Darwin's thought, which is also an extended defense of its coherence and methodological integration, Michael Ghiselin attempts to distinguish between Darwin's "speculations and his definitive view." He denies that Darwin "adopted the inheritance of acquired characteristics because it overcomes the problem of blending." In a complex argument in which he insists on the importance of understanding Darwin's "theoretical constructs," Ghiselin maintains that Darwin's invocation of the inheritance of acquired characteristics was of a different order than that of the theories that have held up: "his ideas about correlated growth and variation, and about latency and throwbacks, which derive from developmental hypotheses supported by an enormous quantity of empirical data, retain much validity even today. By contrast, the idea of blending resulted from a superficial comparison; it was never supported by truly critical tests, and it was founded on no theory. Likewise the effects of use and disuse are not implicit in any theory, were never subjected to rigorous testing, and often occur in Darwin's writings as merely plausible reasons for phenomena equally well explained by selection." Michael Ghiselin, *The Triumph of the Darwinian Method* (1969; Chicago: University of Chicago Press, 1984), 181–82.

5. For an excellent survey of the state of this question in evolutionary biology, see D. S. Wilson, *Darwin's Cathedral*, 11–17. Wilson argues against the dominant view that group selection does not happen, but cautiously sees it as part of a multilevel process of natural selection.

6. Stephen Jay Gould, "Let's Leave Darwin Out of It," *New York Times*, May 29, 1998, A21.

7. See chapter 1, note 33. Young begins with what amounts to a credo for this kind of history: "Only positivists believe that scientific facts and theories are separate from human meanings and values, and even they, inconsistently, set out to extrapolate human and social conclusions from putatively decontextualized facts" (609).

8. James Moore and Adrian Desmond, "Transgressing Boundaries," *Journal of Victorian Culture* 3 (Spring 1998): 156.

9. Walter Bagehot, *Physics and Politics: Thoughts on the Application of the Principles of 'Natural Selection' and 'Inheritance' to Political Society* (1872; New York: Knopf, 1948), 47.

10. Alvar Allegård. *Darwin and the General Reader: The Reception of Darwin's Theory in the British Periodical Press, 1859–1872* (Chicago: University of Chicago Press, 1990), 332.

11. Greta Jones, *Social Darwinism and English Thought: The Interaction between Biological and Social Theory* (Essex, U.K.: Harvester Press, 1980), xii.

12. For a recent reclamation of Darwin for the left, see Peter Singer, *A Darwinian Left: Politics, Evolution, and Cooperation* (New Haven: Yale University Press, 1999).

13. Martin Fichman, *An Elusive Victorian: The Evolution of Alfred Russel Wallace* (Chicago: University of Chicago Press, 2004), particularly 121–23.

14. *Charles Darwin's Notebooks: 1836–1844*, ed. Paul H. Barrett et al. (Ithaca: Cornell University Press, 1987), 518.

15. For a valuable history of this tradition and an analysis of its validity, see Paul Lawrence Farber, *The Temptations of Evolutionary Ethics* (Berkeley: University of California Press, 1994).

16. T. H. Huxley, *Evolution and Ethics*, ed. James Paradis and George Williams (Princeton: Princeton University Press, 1989), 45.

17. With breathtaking confidence in a formulation that helped provoke the great attacks on sociobiology and the various battles that have continued since then, E. O. Wilson concluded his *Sociobiology* in this way: "When mankind has achieved an ecological steady state, probably by the end of the twenty-first century, the internalization of social evolution will be nearly complete. About this time biology should be at its peak, with the social sciences maturing rapidly. Some historians of science will take issue with this projection, arguing that the accelerating pace of discoveries in these fields implies a more rapid development. But historical precedents have misled us before: the subjects we are talking about are more difficult than physics or chemistry by at least two orders of magnitude . . . sociology is drawing closer each day to cultural anthropology, social psychology, and economics, and will soon merge with them." *Sociobiology*, abr. Ed. (1975; Cambridge, MA: Harvard University Press, 1980), 299. In *Consilience*, Wilson is yet more confident of the ultimate unification of knowledge under sociobiology. He argues for "a united system of knowledge," which "is the surest means of identifying the still unexplored domains of reality." *Consilience: the Unity of Knowledge* (New York: Knopf, 1998), 298.

18. See Richard Lewontin, *The Triple Helix* (Cambridge, MA: Harvard University Press, 2002). Lewontin's basic argument is that, as the press release notes, "we will never fully understand living things if we continue to think of genes, organisms, and environments as separate entities, each with its distinct role to play in the history and operation of organic processes. . . . an organism is a unique consequence of both genes and environment, of both internal and external features. Rejecting the notion that genes determine the organism, which then adapts to the environment, he explains that organisms, influenced in their development by their circumstances, in turn create, modify, and choose the environment in which they live."

19. See Eiseley's essay "Man against the Universe," in *The Star Thrower* (New York: Harcourt Brace, 1978), 207–21. Eiseley compares Darwin unfavorably to Emerson, complaining about Darwin's constant emphasis on struggle, about his polite gentleman's invocation of the "creator," and about the way, though Darwin had "the soul of a romantic," he relied on a thoroughgoing materialistic and mechanistic interpretation of nature.

20. Steven Weinberg, *Dreams of a Final Theory: The Scientist's Search for the Ultimate Order of Nature* (New York: Vintage Books, 1993), 54.

21. John Dupré, *The Disorder of Things: The Metaphysical Foundations of the Disunity of Science* (Cambridge, MA: Harvard University Press, 1993), 87–167.

22. In the introduction to his textbook on evolutionary psychology, David Buss argues that the critique of sociobiology is mistaken in charging that it is committed to genetic determinism: "Much of the resistance to applying evolutionary theory to the understanding of human behavior stems from the misconception that evolutionary theory implies genetic determinism. Contrary to this misunderstanding, evolutionary theory in fact represents a truly interactionist framework." *Evolutionary Psychology: The New Science of Mind* (Needham Heights, MA: Allyn and Bacon, 1999), 19.

23. In her important study of the debates over sociobiology, Ullica Segerstråle suggests a very complicated answer to the question of the ideological work of sociobiology. For the critics of sociobiology, she says at the outset, its true meaning could be found in a "moral reading." "For them the political truth of sociobiology was obvious. For the sociobiologists, it was not" (2). On the whole, Segerstråle's history suggests the ideological innocence of the sociobiologists, who were seeking "a new way of understanding evolution." She recognizes that science is often driven by moral impulses but holds to the view that (quoting Maynard Smith), "scientific theories say nothing about what is right, but only about what is possible." *Defenders of the Truth: The Battle for Science in the Sociobiology Debate and Beyond* (Oxford: Oxford University Press, 2000), 403.

24. Although Darwin's ideas have often been assimilated to religious positions, Darwin explicitly rejected Asa Gray's effort to do that and to demonstrate that the theory led inevitably to improvement. "However much we may wish it, we can hardly follow Professor Asa Gray in his belief 'that variation has been led along certain beneficial lines,' like a stream 'along definite and useful lines of irrigation.' If we assume that each particular variation was from the beginning of all time preordained, then that plasticity of organisation, which leads to many injurious deviations of structure, as well as the redundant power of reproduction which inevitably leads to a struggle for existence, and, as a consequence, the natural selection or survival of the fittest, must appear to us superfluous laws of nature." Charles Darwin, *The Variation of Plants and Animals under Domestication*, 2 vols. (1868, 1875; New York: Appleton, 1900), 2:415. For an analysis of the whole passage to which this belongs, see chapter 5, below.

25. Janet Browne, "I Could Have Retched All Night: Charles Darwin and His Body," in *Science Incarnate: Historical Embodiments of Natural Knowledge*, ed. Christopher Lawrence and Steven Shapin (Chicago: University of Chicago Press, 1998), 240–87.

26. Barbara Herrnstein Smith, *Contingencies of Value: Alternative Perspectives for Literary Theory* (Cambridge, MA: Harvard University Press, 1988), 47–49.

27. Henry Plotkin, *Darwin Machines and the Nature of Knowledge* (Cambridge, MA: Harvard University Press, 1994), 48–57.

28. John Dewey, "The Influence of Darwin on Philosophy," in *Darwin*, Norton Critical Edition, ed. Philip Appleman (New York: W. W. Norton, 1970), 393.

29. It is Dennett who translates Darwin's metaphor into "algorithm": see *Darwin's Dangerous Idea* (New York: Simon and Schuster, 1995), 50, but also my discussion of the word in chapter 7, below. Interestingly, Dennett begins this argument with an epigraph from the *Origin*: "What limit can be put to this power, acting during long ages and rigidly scrutinising the whole constitution, structure, and habits of each creature—devouring the good and rejecting the bad? I can see no limit to this power, in slowly and beautifully adapting each to the most complex relations of life."

30. Niles Eldredge, *Reinventing Darwin: The Great Evolutionary Debate* (London: Orion Books, 1995), 76.

31. Oscar Kenshur, *Dilemmas of Enlightenment: Studies in the Rhetoric and Logic of Ideology* (Berkeley: University of California Press, 1993), 7–8. For my own initial discussion of this point, see *Dying to Know*, 17.

32. Paul Crook, *Darwinism, War, and History* (Cambridge: Cambridge University Press, 1994), 15.

33. Richard Dawkins, *A Devil's Chaplain: Reflections on Hope, Lies, Science and Love* (Boston: Houghton, Mifflin, 2003), 10–11.

CHAPTER 3

1. Janet Oppenheim, *The Other World: Spiritualism and Psychical Research in England, 1850–1914* (Cambridge: Cambridge University Press, 1985), 268.

2. Steve Jones, "In the Genetic Toyshop," *New York Review of Books* 45 (April 23, 1998): 15.

3. Quoted in D. R. Oldroyd, *Darwinian Impacts: An Introduction to the Darwinian Revolution* (Atlantic Highlands, NJ: Humanities Press, 1980), 36.

4. Most recently and interestingly by D. S. Wilson in *Darwin's Cathedral*. But Wilson's argument is quite self-consciously secular and Darwinian in the sense that it argues for the adaptive function of religion, rather than for any transcendental authority.

5. For an interesting survey of the problem, see Gregory Radick, "Is the Theory of Natural Selection Independent of Its History?" (Hodge and Radick, 143–67). According to Radick, the claim that Darwin developed the theory as he did simply because it is true is a "non-starter." While Radick makes it clear how much the theory's particular shape was informed by elements in Darwin's culture, he does not talk at all about whether the theory might be used legitimately, after the fact, with different cultural inflections.

6. Peter Bowler, *The Eclipse of Darwinism: Anti-Darwinian Evolution Theories in the Decades around 1900* (Baltimore: Johns Hopkins University Press, 1983).

7. Theodore M. Porter, whose recent biography of Pearson lays out clearly Pearson's relation to Darwinian thought and the connection of that with his socialism, writes that Pearson was angered by the popular success of Kidd's

Social Evolution since he did not believe that Kidd was a scientist: "He had no truck with Kidd's individualistic politics, his doctrine that socialism meant biological degeneration. Kidd called for a biological religion as a kind of opiate, to reconcile the unfit to the hardships and restrictions that were unavoidable if natural selection was to continue to operate." *Karl Pearson: The Scientific Life in a Statistical Age* (Princeton: Princeton University Press, 2004), 202.

8. Benjamin Kidd, *Social Evolution* (New York: Macmillan and Co., 1894), 26.

9. Karl Pearson, *The Grammar of Science* (London: Adam and Charles Black, 1900), 24.

10. Richards, who holds that Darwin's theory is pervaded by romantic ideas and sentiments, argues that "altruism" was always part of Darwin's way of looking at the world, whatever has been made of his theory since (see 546–47, 549–51). I will want to develop this idea in later chapters, but it is crucial to recognize that the current sense of the incompatibility of altruism with natural selection is not consistent with Darwin's views, with his rejection of utilitarian ethics (despite regular accusations that his ethics are strictly utilitarian), or with his metaphorical imagination of the work of natural selection.

11. This passage has been crucial to most attempts to read Darwin out of the normal framework of social Darwinist insistence on competition and violence. See the excellent discussion by Crook (15–20).

12. I use Kidd's quotations with his italics rather than Darwin's original. The quotations are from the sixth edition of the *Origin*.

13. "The concept of group selection has been strongly criticized and virtually dismissed by most scientists. Assaults by George C. Williams, Richard Dawkins, and William Hamilton have nearly destroyed the notion" (Farber, 152).

14. Wilson is one of the most interesting and successful evolutionary biologists to reaffirm the possibility of group as opposed to individual selection, a dominant view in Kidd's time but, as I have indicated, rejected totally in the mid-twentieth century. With Elliot Sober, Wilson wrote *Unto Others: The Evolution and Psychology of Unselfish Behavior* (Cambridge, MA: Harvard University Press,1998), a careful and sustained argument for the possibility of group selection.

15. Benjamin Kidd, *Principles of Western Civilization* (New York: Grosset and Dunlap, 1902), 463–64.

16. Karl Pearson, "Socialism and Natural Selection," in *The Chances of Death and Other Studies in Evolution* (London: Edwin Arnold, 1897), 1:103.

17. While in his 1894 essay, "Socialism and Natural Selection," Pearson strongly emphasizes the socialist side of his scientific conclusions and thus makes compassion scientifically rational, his fully developed theory explains the need for socialism by way of emphasis on competition among nations and races. In an address to the Literary and Philosophical Society of Newcastle, Pearson empahsizes not socialism but national struggle. In the wake of the Boer War, he proposes a scientific rather than an emotional or moralistic analysis and indicates that one of the lessons of that war is that "the struggle for existence among nations" will be settled not by size or brute strength but by the wisdom and unity of the combatant nations. Wisdom requires unity, and thus it is necessary to keep even the inferior members of society happy—ergo,

socialism. But though socialism is thus scientifically enjoined, so too is the determination to crush inferior races, and so too is the need to ensure that the inferior members of one's own society do not outbreed the superior ones—hence eugenics. Here is how he makes the case for socialism in the address: "We, as a nation, cannot survive in the struggle for existence if we allow class distinctions to permanently endow the brainless and to push them into posts of national responsibility. The true statesman has to limit the internal struggle of the community in order to make it strong for the external struggle. We must reward ability, we must pay for brains, we must give larger advantage to physique but we must not do that at a rate which renders the lot of the mediocre an unhappy one. We must foster exceptional brains and physique for national purposes; but, however useful prize-cattle may be, they are not bred for their own sake, but as a step towards the improvement of the whole herd"(52).

In the light of history, the talk is a frightening one. Here, for example, is a characteristic passage about the races: "If you bring the white man into contact with the black, you too often suspend the very process of natural selection on which the evolution of a higher type depends. You get superior and inferior races living on the same soil, and that co-existence is demoralizing for both. They naturally sink into the position of master and servant, if not admittedly or covertly into that of slave-owner and slave. Frequently they intercross, and if the bad stock be raised the good is lowered"(20). See Karl Pearson, *National Life from the Standpoint of Science* (London: Adam and Charles Black, 1900).

18. Peter Kropotkin, *Mutual Aid* (New York: Black Rose Books, 1989), xxxvii.

CHAPTER 4

1. Richard Lewontin, *It Ain't Necessarily So: The Dream of the Human Genome and Other Illusions* (New York: New York Review of Books, 2000), 52.

2. Steven Pinker, *The Blank Slate: The Modern Denial of Human Nature* (New York: Viking, 2002), 32.

3. Lewontin, who certainly is not enamored of the reductionist arguments of Richard Dawkins, makes an almost Dawkinsian point about "God": "Nature is at constant risk before an all-powerful God who at any moment can rupture natural relations. For sufficient reason, He may just decide to stop the sun, even if He hasn't done so yet. Science cannot coexist with such a God. If, on the other hand, God *cannot* intervene, he is not God; he is an irrelevancy. By failing to confront this problem, biologists and philosophers may make Unitarians and agnostics feel that ontological pluralism is a happy solution, but they haven't fooled any fundamentalists, who know better" (*It Aint Necessarily So*, 51).

4. In his excellent study of Victorian psychology, Rick Rylance points out that in late-Victorian England, however, it was Spencer rather than Darwin that most influenced the movements of psychology, readers finding it in fact difficult to disentangle Darwin's from Spencer's position, which emphasized the inheritance of acquired characteristics. Rylance argues that Darwin's real influence on psychology had to wait until later, and in fact is getting its fullest application in our own time. In an argument to the point of this chapter,

Rylance notes that "in some respects, Darwin became painted with Spencer's political brush, and 'Social Darwinism' became the usual term for the biologization of nineteenth-century political economy." *Victorian Psychology and British Culture: 1850–1880* (Oxford: Oxford University Press, 2000), 224.

5. Roger Smith, *The Human Sciences* (New York: W. W. Norton, 1997), 453.

6. Charles Darwin, *The Expression of the Emotions in Man and Animals*, 3d. ed. (1872; New York: Oxford University Press, 1998), 360.

7. Cited in *Darwin's Cathedral*, 67.

8. Dennett rejects what he calls *greedy reductionism*, as opposed to the good kind, "non-question-begging science without any cheating by embracing mysteries or miracles at the outset." Greedy reductionism, he explains, rushes too hastily to "fasten everything securely and neatly to the foundation." He claims that "Darwin's dangerous idea is reductionism incarnate, promising to unite and explain just about everything in one magnificent vision" (82).

9. Richard Dawkins, *The Blind Watchmaker* (New York: W. W. Norton, 1986), 13.

10. John Dupré, *Human Nature and the Limits of Science* (Oxford: Oxford University Press, 2001), 72.

11. Richard Lewontin, Steven Rose, and Leon J. Kamin, *Not in Our Genes* (New York: Pantheon Books, 1984), 6.

12. John Dupré, *The Disorder of Things: Metaphysical Foundations of the Disunity of Science* (Cambridge, MA: Harvard University Press, 1993). Dupré takes up three aspects of the argument for unity: reductionism, essentialism, and causality, or determinism, and attempts to show that the denial of unity does not at all call into question "the entire scientific edifice."

13. Segerstråle, 25. Segerstråle quotes from Wilson's autobiography, *Naturalist*.

14. Auguste Comte, *Auguste Comte and Positivism: The Essential Writings* (Chicago: University of Chicago Press, 1975), 72.

15. William Whewell, *Theory of Scientific Method*, ed. Robert E. Butts (Indianapolis: Hackett, 1989), 138.

16. Laura Dassow Walls, unpublished address to the History of Science Society, November 1999.

17. Robert E. Butts describes consilience simply, and in ways that are to a certain extent consistent with Wilson's ambitious adaptation. "Whewell," says Butts, "thought that the best test of any scientific explanation or theory is what he called *consilience*. A consilience of inductions explains data of a kind different from those it was initially introduced to explain . . . a proper consilience of inductions takes place when data of a kind different from the deductive expectations of a theory are seen to be of the same kind as those initially colligated" (Whewell, 28).

18. Spencer's system entailed a belief in what he called "universal progress." In Spencer's voluminous work the ramifications of this process are of course applied directly to contemporary social conditions. Nevertheless, it is worth being reminded of the *way* Spencer talked on these matters. In his essay "Universal Progress," he wrote: "Should the Nebular Hypothesis ever be established, then it will become manifest that the Universe at large, like every organism, was once homogeneous; that as a whole, and in every detail, it has

unceasingly advanced towards greater heterogeneity; and that its heterogeneity is still increasing. It will be seen that as in each event of to-day, so from the beginning, the decomposition of every expended force into several forces has been perpetually producing a higher complication; that the increase of heterogeneity so brought about is still going on, and must continue to go on; and that thus Progress is not an accident, not a thing within human control, but a beneficent necessity." *Illustrations of Universal Progress: A Series of Discussions* (New York: Appleton and Co., 1864), 58. Spencer's faith in progress, the movement from the homogenous to the heterogeneous, was absolute, but unlike Wilson he sustained a deep sense of the ultimate "impenetrable mystery." In some versions of contemporary science, that mystery has been dispelled.

19. Carl Degler, *In Search of Human Nature: The Decline and Revival of Human Nature in American Social Thought* (New York: Oxford University Press, 1991), 318–19.

20. John Alcock, *The Triumph of Sociobiology* (New York: Oxford University Press, 2001), 218.

CHAPTER 5

1. Arthur Balfour, *The Foundations of Belief* (1894; New York: Longmans, Green, and Co., 1895), 15.

2. Both Mallock and Balfour, in slightly different ways, reject the biographical argument. Whatever their differences, both agree that the personal goodness of the atheists, which neither author denies, is derived, without acknowledgment, from religion rather than from their naturalist views. Mallock believes that the atheists can be good only insofar as they borrow their morality from religious traditions. In a way he makes a powerful case for the fact of disenchanting effects of science, arguing in effect that there can be no meaning or value in a strictly natural world. Compare Balfour's comments on this issue. Many naturalist thinkers are in fact exemplary people morally, but "Their spiritual life is parasitic; it is sheltered by convictions which belong, not to them, but to the society of which they take no share. And when those convictions decay, and those processes come to an end, the alien life which they have maintained can scarce be expected to outlast them" (83).

3. Charles Darwin, *Diary of the Voyage of the H.M.S. Beagle*, ed. Nora Barlow, in *The Works of Charles Darwin*, ed. Paul H. Barrett and R. B. Freeman (London: William Pickering, 1986), 42.

4. See Donald Fleming, "Charles Darwin, the Anaesthetic Man," *Victorian Studies* 4 (1961): 219–36.

5. For a fascinating discussion of stochastic processes and of the possibility they offer for new ways to think about nature and connect with it, see Gregory Bateson, *Mind and Nature: A Necessary Unity* (New York: E. P. Dutton, 1979).

6. As many commentators have noted, Darwin is not entirely consistent on this point. In the *Origin*, he is usually careful to indicate that evolutionary change is responsive to new environmental conditions, and there is no reason for those changes to move always on the lines of progress. "Perfection," as most modern commentators understand Darwin's position, is only perfection within a particular environmental context; any change in the context makes the

perfect organism vulnerable and imperfect. In the *Descent*, dealing with the development of humans, Darwin implies that there has been a progress from the lowest to the highest, culminating in civilized humans, who have both an ethical and aesthetic sense. Richards, in *The Romantic Conception of Life*, makes a strong case for Darwin's commitment to progress, demonstrating that at least in his earlier phases, his work reflected progressivist thinking. There is a stunning sentence in the autobiography, restored by Nora Barlow, that gives another ambivalent Darwinian view on the issue: "Believing as I do that man in the distant future will be a far more perfect creature than he now is, it is an intolerable thought that he and all other sentient beings are doomed to complete annihilation after such long-continued slow progress" (92).

7. Near the end of *The Variation of Animals and Plants under Domestication*, Darwin elaborates a powerful case against the idea that every detail of the evolutionary process must be fully explained before the theory can be accepted. Using the analogy of a building built from random stones, as the builder carefully chose stones of different shapes for different purposes, he argues that, "if it were explained to a savage utterly ignorant of the art of building, how the edifice had been raised stone upon stone, and why wedge-formed fragments were used for the arches, flat stones for the roof, &c.; and if the use of each part and of the whole building were pointed out, it would be unreasonable if he declared that nothing had been made clear to him, because the precise cause of the shape of each fragment could not be told. But this is a nearly parallel case with the objection that selection explains nothing, because we know not the cause of each individual difference in the structure of each being" (2:414). He goes on to challenge the idea that (although a long, elaborate historical investigation might indeed explain how each rock got its shape) "the Creator intentionally ordered, if we use the words in any ordinary sense, that certain fragments of rock should assume certain shapes so that the builder might erect his edifice." Similarly, "no shadow of reason can be assigned for the belief that variations, alike in nature and the result of the same general laws, which have been the groundwork through natural selection of the formation of the most perfectly adapted animals in the world, man included, were intentionally and specially guided" (2:415).

8. Letter to Joseph Hooker, January 11, 1844 *Correspondence*, 3:2.

9. Francis Darwin, ed., *The Life and Letters of Charles Darwin*, 3 vols. (London: John Murray 1887), 3:75.

10. Philip Sloan, "The Making of a Philosophical Naturalist," in Hodge and Radick, 30.

11. James Paradis, "Darwin and Landscape," in *Victorian Science and Victorian Values: Literary Perspectives*, ed. James Paradis and Thomas Postlethwaite (New York: New York Academy of Sciences, 1981), 85. Paradis goes on to point out that "Darwin traced the historical path from poetic to scientific nature in his M transmutation Notebook in 1838" (101). I would only want to argue that the "path" was not a simple "from/to," that the poetry and the science were mutually involved with each other.

12. In a fascinating recent essay, Christoph Irmscher argues that Darwin's last book, on vegetable mold and worms, is a kind of summation of the bleak vision that Darwin had of his own life, as well as of the life of the world. The

worms, mechanically ingesting and excreting the earth, without power to respond to the aesthetic (the music with which Darwin had his family try to entertain them), are like Darwin, ingesting and excreting facts. The argument is a strong one, but as with most commentaries on Darwin, it makes its case for the bleak side of the Darwinian vision, whereas *Vegetable Mould* is a strangely exhilarating book, in which Darwin shows that the worms *do* respond. The book demonstrates in its minute attention to almost the lowest of living things a deep sense of wonder at the creativity of the world. See Christoph Irmscher, "Darwin's Beard," in *LIT*, 2004, 87–105.

13. Howard Gruber, "Going the Limit: Toward the Construction of Darwin's Theory (1832–1839)," in Kohn, 32.

14. Rebecca Stott, *Darwin and the Barnacle: The Story of One Tiny Creature and History's Most Spectacular Scientific Breakthrough* (London: Faber and Faber, 2003).

15. Charles Darwin, *The Formation of Vegetable Mould through the Action of Worms with Observations on Their Habits* (London: John Murray, 1883), 85.

16. Charles Darwin, *Voyage of the* Beagle, ed. Janet Browne and Michael Neve (1839; London: Penguin Books, 1989), 41.

17. *The Works of John Ruskin: Library Edition*, ed. E. T. Cook and Alexander Wedderburn (London: George Allen, 1904), 4:128.

18. Desmond and Moore discuss the way "Family inbreeding had long worried him" (*Charles Darwin*, 447). They point out that by 1857, while Darwin was laboring over his "big book" (what the *Origin* might have been had not Wallace also discovered the theory of "natural selection"), two of his ten children had died from natural causes, and there were "ominous" signs for the rest: "Charles believed that the main problem was hereditary: that his own constitutional weakness had been passed on, accentuated by Emma's Wedgwood blood. The struggle for existence had already set in" (447). According to Desmond and Moore, Darwin believed that "There was no escaping Nature's ruthless scythe, and no virtue in the attempt" (448).

19. John Bowlby, *Charles Darwin: A New Life* (New York: W. W. Norton, 1990), 23.

20. James Moore, "Of Love and Death: Why Darwin 'Gave Up Christianity,'" in *History, Humanity, and Evolution: Essays for John C. Greene*, ed. James Moore (New York: Cambridge University Press, 1989). In Randal Keynes, *Darwin, His Daughter and Evolution* (New York: Riverhead Books, 2002), there are two extensive chapters devoted to Annie's illness, the details of the attempted cure, and to Darwin's thoughts, in particular, about religion; see chapters 9–11, in particular. Rebecca Stott writes sensitively about the event and juxtaposes the prose Darwin used in his barnacle work, on which he was laboring at the time of Annie's death, and the language in his memorial for Annie (168ff.). She discusses at length Etty's religious worries about her sister's death, and speculates on Darwin's response to them: "Perhaps [Darwin] wanted to say what he was beginning to feel for himself: that these were all 'notions of the cave' and that after death there was nothing—no God waiting to scour Annie's record book to decide whether she would be consigned to heaven or hell. She had only to be strong for life; and she hadn't been strong enough" (171).

21. For a more detailed and quite moving representation of Darwin's relation to Annie's illness and death, see Stott, 154–71. Stott discusses Darwin's earlier habit of keeping a diary of his babies' behavior, the six-year hiatus before he took up again, in tracing Annie's condition, a diary of health. She points out also how Darwin continued reading agnostic and free-thinking material during that time, and she fills in the context of his work on barnacles, which he interrupted to follow Annie to the sanitarium. The continuity between his scientific relation to those organisms and his deeply felt registration of Annie's condition is striking.

22. William Paley, *Natural Theology: or, Evidences of the Existence and Attributes of the Deity, Collected from the Appearances of Nature* (Boston: Gould and Lincoln, 1860), 287.

23. This change appeared first in the third edition of the *Origin* and remained in all succeeding ones.

24. See all of chapter 44, "An Agnostic in the Abbey," which narrates the movement to inter Darwin in the Abbey, and James Moore, "Charles Darwin Lies in Westminster Abbey," *BJLS* 17 (1982): 97–113.

25. "Charles turned gratefully back to his routines and to the busy life of Down, seeking to fill the gap that Annie had left behind, keeping busy. As he walked around the village, the men from the local Friendly Club who played cricket in his meadow, the carpenters and blacksmiths who had worked on his house and the village shopkeepers passed on their condolences. At home the house was full of new life—the two little boys, Francis and Lenny, for whom Annie was only a passing shadow, no more important than the nursemaids who chased them, played in the garden as before, running after butterflies and catching ladybirds. Charles lay out under the big trees, feeling the warmth of the sun on his face, while the little boys climbed over him, pretending that he was a mountain bear, running their hand through the thick hairs on his chest and arms. He heard their shouts and laughs when he sat in his study. The frame of the study window was filled with startling green—the fresh leaves of lime trees. There were letters to answer and a book to finish" (Stott, 173).

CHAPTER 6

1. Jonathan Smith, comment on early version of this chapter presented at MLA Conference, December 2000.

2. "Darwin's Notes on Marriage," July 1838 (*Correspondence*, 2:444–45).

3. Angelique Richardson, in a series of essays, has done extremely valuable work on the question of sexual selection in relation to Hardy and several interesting late-nineteenth-century novelists. See, for example, "Some Science Underlies All Art: The Dramatization of Sexual Selection and Racial Biology in Thomas Hardy's *A Pair of Blue Eyes* and *The Well-Beloved*," *Journal of Victorian Culture* 3/2 (1998): 302–28. Richardson is highly sensitive to questions of gender and race throughout, but her readings of Darwin and his adaptation by the writers in question are also sensitive to the complexity of his arguments.

4. Gillian Beer, *Darwin's Plots: Evolutionary Narrative in Darwin, George Eliot and Nineteenth-Century Fiction* (London: Routledge, Kegan Paul, 1983), 213. A revised edition was published by Cambridge University Press in 2000.

5. For a careful early analysis of the relation of Darwin's thinking to contemporary political economy, see Schweber, "Darwin and the Political Economists."

6. Rosemary Jann, "Darwin and the Anthropologists: Sexual Selection and Its Discontents," *Victorian Studies* 37 (1994): 286–306.

7. Eveleen Richards, "Darwin and the Descent of Woman," in *The Wider Domain of Evolutionary Thought*, ed. David Oldroyd and Ian Langham (Reidel: Dordrecht, 1983), 61.

8. Fiona Erskine, "*The Origin of Species* and the Science of Female Inferiority," in *Charles Darwin's Origin of Species: New Interdisciplinary Essays*, ed. David Amigoni (New York: St. Martin's Press, 1995), 95–121.

9. For an interesting discussion of these two writers' relation to sexual selection, see Helena Cronin, *The Ant and the Peacock* (Cambridge: Cambridge University Press, 1991), esp. 172.

10. Gertrude Himmelfarb, *Darwin and the Darwinian Revolution* (London: Chatto and Windus, 1959).

11. Cynthia Russett, *Sexual Science: The Victorian Construction of Womanhood* (Cambridge, MA: Harvard University Press, 1989), 92.

12. Oscar Kenshur, *Dilemmas of Enlightenment: Studies in the Rhetoric and Logic of Ideology* (Berkeley: University of California Press, 1993), chapter 1, "Ideological Essentialism."

13. In *Philosophy and Social Hope* (London: Penguin, 1999), Richard Rorty similarly argues that large philosophical theories are not linked to particular political positions. "Both the orthodox and the postmoderns still want a tight connection between people's politics and their view on large theoretic (theological, metaphysical, epistemological, metaphilosophical) matters" (18). "People on the left keep hoping for a philosophical view which cannot be used by the political right, one which will lend itself only to good causes. But there never will be such a view; any philosophical view is a tool which can be used by many different hands" (23).

14. Frederick Burkhardt, "Darwin on Animal Behavior and Evolution," in Kohn, 357.

15. *The Triumph of Darwinian Method*, with a new preface (1969; Chicago: University of Chicago Press, 1984).

16. See, in particular, Ospovat, 60–61.

17. The phrase appears in the second edition of the *Descent*, ed. Desmond and Moore, 261.

CHAPTER 7

1. "Genetic determinism" has been a red flag in the conflicts over sociobiology and evolutionary psychology. In one sense, as Philip Kitcher, *Vaulting Ambition* (Cambridge, MA: MIT Press, 1985) argues, it is a red herring, since nobody, on either side of the debates, believes in an absolute genetic determinism: "Pop sociobiologists and their opponents agree that genes and environment together determine phenotype and that is the end of the matter. . . . Nobody believes in the iron hand of the gene, and nobody believes in the blank mind" (24). But Kitcher elaborates in a far more complex way the nature of the serious

dispute that remains beyond the flag of "genetic determinism," one that re-volves around the degree to which environmental factors can be determined, and the degree to which "a particular reduction of the environmental variables effectively represents" the "complex mapping" of environmental factors onto behavior (26). The reductionist mode by which sociobiologists tend to identify those environmental factors allows me the simplification here of "biological determinism."

2. Ernst Haeckel conveniently summarizes (his own version of) the history of the development of the idea of evolution in *Die Welträthsel* (1899). He de-scribes himself as the first to take up the responsibility, suggested by Darwin's work, to reform the zoological and botanical "*system*." Against the traditional view that "evolution" applied only to individual organisms, he "established the opposite view, that this history of the embryo (ontogeny) must be com-pleted by a second, equally valuable, and closely connected branch of thought—the history of the race (phylogeny)." *The Riddle of the Universe*, trans. Joseph McCabe (1899; New York: Harper Brothers, 1900), 80.

3. Richards goes on to indicate what it is that makes a historical representa-tion ideological: "first, the historical account employs an interpretative framework or set of assumptions that are covert and neither justified nor ar-gued for in the account; second, the framework or assumptions express the shared values and position of a particular community rather than the idio-syncratic view of the historian; third, the main function of the framework or assumptions is to justify the shared values and position rather than to realize the principal value of recovering the past; and finally, the historian's inter-pretations and arguments serve chiefly to justify the framework and thus the values" (175).

4. Frances Cobbe's hostility to Darwin's worldview was made explicit in many places, not only in their well-known clash over vivisection, but also in her deep aversion to the moral implications of his work. In *Darwinism in Morals* (1872), she devotes a long essay to the problems with Darwin's natura-listic arguments. In their introduction to a recent edition of *The Descent of Man*, Adrian Desmond and James Moore quote Cobbe describing Darwin as "a man who has . . . unconsciously attributed his own abnormally generous and placa-ble nature to the rest of his species, and then theorized as if the world were made of Darwins" (lvii). Darwin, the nice guy, is separated off from his theory; I am trying at least partially, to reconnect them.

5. I conduct a parallel discussion of this issue in my *Dying to Know*; see chap-ter 4.

6. Evelyn Fox Keller, *A Feeling for the Organism: The Life and Work of Barbara McClintock* (New York: W. H. Freeman, 1983), 198.

7. For a counternarrative, see Jonathan Smith, *Fact and Feeling: Baconian Sci-ence and the Nineteenth-Century Imagination* (Madison: University of Wisconsin Press, 1994).

8. John Tyndall, "The Belfast Address," *The Victorian Web*, digitized by John van Wyhe, 44–45. http://www.victorianweb.org/science/science_texts/belfast.html.

9. Desmond and Moore, in their biography of Darwin, point to an example of how Darwin somewhat deviously "cajoled" and "baited" Huxley into writing a

favorable review of Darwin's barnacle book by first offering Huxley the opportunity to examine some specimens of Ascidae that Darwin had. It will "give me some trouble" to make them available, Darwin says, "but it would give me *real pleasure* should you wish to examine them." Desmond and Moore note, though, that Huxley seems already to have indicated to Darwin that he would be quite pleased to review the book. Aware of what his suggestion might seem like, Darwin apologizes for "the length and egotistical character of this note." Janet Browne discusses the same letter in a slightly different tone, but it is reasonable to argue that Darwin was not above exploiting his own kindness to evoke support from others. See Desmond and Moore, *Charles Darwin*, 406; *Correspondence*, 5:130; Browne, *Voyaging*, 507.

10. Donald Fleming, "Charles Darwin, the Anaesthetic Man," *Victorian Studies* 4 (1961): 219–36.

11. Stott draws the passage quoted here from Darwin's *A Monograph of the Sub-Class Cirripedia* (London: Ray Society, 1851), 292–93.

12. Carolyn Merchant, *The Death of Nature: Women, Ecology, and the Scientific Revolution* (New York: Harper Collins, 1989). See esp. 227–35.

13. Mary Louise Pratt, *Imperial Eyes: Studies in Travel Writing and Transculturation* (New York: Routledge, 1992), 30.

14. John Durant, "The Ascent of Nature in Darwin's *Descent of Man*, in Kohn, 287–88.

15. As the editors of volume 4 of *The Correspondence of Charles Darwin* explain, "it was Darwin's personal experience of fatherhood that was central to his formulation of the questions he was to pursue regarding the nature of the expression of emotions" (410). Appendix 3 of that volume includes all his notebook entries about his children, not only William, of course. The notes run from 1839, when William was born, until 1856. Some of the entries in the notebooks are by Emma Darwin.

16. Reprinted in Howard Gruber, *Darwin on Man: A Psychological Study of Scientific Creativity* (Chicago: University of Chicago Press, 1981), 464–74.

17. Charles Darwin, *The Origin of Species: A Variorum Text*, ed. Morse Peckham (Philadelphia: University of Pennsylvania Press, 1959), 165.

EPILOGUE

1. George Charpak and Roland Omnes, *Be Wise, Become Prophets* (2004), quoted from an excerpt published in *La repubblica*, October 20, 2004; my translation.

2. See Callum Brown, *The Death of Christian Britain* (New York: Routledge, 2001), in particular, 193–95.

3. I must thank my good friend and fellow birder and Darwinian, Christopher Herbert, for reminding me of this fact.

4. *New York Times*, September 13, 2005, Science section.

5. Joseph Vining, *The Song Sparrow and the Child: Claims of Science and Humanity* (Notre Dame, IN: University of Notre Dame Press, 2004), 142. This remarkable book is a sustained and learned argument against all of what Vining calls "totalizing theory," and seeks, in a moving and credible way, to locate "spirit" in the natural world, to resist pure scientism without resisting science.

6. In a recent conversation with me, David Albert, the distinguished philosopher of science, argued that if it were possible to find a totally deterministic and materialist explanation of the full range of human consciousness and action, enchantment would be impossible. My argument, that scientists find the world enchanting even as they explain its intimate workings, would not hold in such conditions. Nothing would have any "meaning" because all phenomena would be the products of a sort of mindless algorithm, the laws of nature working themselves out absolutely no matter how much our minds deceived us into believing that we were making choices, responding to desires, etc. Of course, Albert's theoretical notion of a total explanation of everything in naturalistic terms remains only a fantasy, but it is worth speculating about how the full move from mystery, to puzzle, to resolution would affect the moral and cultural problems I am considering throughout this book.

7. I use the summary here of Andrew Brown, *The Darwin Wars: How Stupid Genes Became Selfish Gods* (New York: Simon and Schuster, 1999), 2–3.

8. Samuel Butler, *Unconscious Memory* (1880; London: Jonathan Cape, 1922), 175. His three books, including *Unconscious Memory*, *Life and Habit* (1877), and *Luck or Cunning* (1887), are extended arguments for evolution and against Darwin, brilliant and insightful even as they formulate a position that has been thoroughly rejected by twentieth-century science.

9. The most important book in the development of the argument for design, which has become the most recent cover for creationism, and a particularly effective one, is Michael J. Behe, *Darwin's Black Box* (New York: Touchstone Books, 1998). Behe argues that the history of modern science is the history of the discovery of increasing complexity in nature: "The simplicity that was once expected to be the foundation of life has proven to be a phantom; instead, systems of horrendous, irreducible complexity inhabit the cell. The resulting realization that life was designed by an intelligence is a shock to us in the twentieth century who have gotten used to thinking of life as the result of simple natural laws" (252). Of course, it has been a long time since anyone thought such laws were "simple," but the assumption that, having found "complexity," we are necessarily driven to the belief that life was "designed by an intelligence" is a further manifestation of a very widespread natural-theological belief that the model for any design is human intelligence. Of course, implicitly, Behe is implying a divine intelligence that is too smart by a long way for human intelligence. It is the dogged believer's version of the astonishment and awe Darwin so frequently expressed when confronted with complex phenomena: "What shall we say," Darwin asks, "to so marvelous an instinct as that which leads the bees to make cells which have practically anticipated the discoveries of profound mathematicians?" (172). The same problem—and Darwin spends a chapter trying, with some success, to explain what indeed we might say, and convincingly constructs an explanation that makes the bees, in their way, outstanding mathematicians!

10. John Stuart Mill, *Autobiography and Other Writings*, ed. Jack Stillinger (Boston: Houghton Mifflin, 1969), 91–92.

11. Quoted in Kate Flint, *The Victorians and the Visual Imagination* (Oxford: Oxford University Press, 2000), 62, from W. B. Carpenter, *The Microscope and Its Revelations* (1856).

INDEX